POLICIES AND INVESTMENTS TO ADDRESS CLIMATE CHANGE AND AIR QUALITY IN THE BEIJING–TIANJIN–HEBEI REGION

DECEMBER 2022

ASIAN DEVELOPMENT BANK

© 2022 Asian Development Bank
6 ADB Avenue, Mandaluyong City, 1550 Metro Manila, Philippines
Tel +63 2 8632 4444; Fax +63 2 8636 2444
www.adb.org

Some rights reserved. Published in 2022.

ISBN 978-92-9269-918-5 (print); 978-92-9269-919-2 (electronic); 978-92-9269-920-8 (ebook)
Publication Stock No. TCS220551
DOI: http://dx.doi.org/10.22617/TCS220551

The views expressed in this publication are those of the authors and do not necessarily reflect the views and policies of the Asian Development Bank (ADB) or its Board of Governors or the governments they represent.

ADB does not guarantee the accuracy of the data included in this publication and accepts no responsibility for any consequence of their use. The mention of specific companies or products of manufacturers does not imply that they are endorsed or recommended by ADB in preference to others of a similar nature that are not mentioned.

By making any designation of or reference to a particular territory or geographic area, or by using the term "country" in this document, ADB does not intend to make any judgments as to the legal or other status of any territory or area.

Please contact pubsmarketing@adb.org if you have questions or comments with respect to content, or if you wish to obtain copyright permission for your intended use that does not fall within these terms, or for permission to use the ADB logo.

Corrigenda to ADB publications may be found at http://www.adb.org/publications/corrigenda.

Notes:
In this publication, "$" refers to United States dollars.
ADB recognizes "China" as the People's Republic of China and "United States of America" as the United States.

Cover design by Claudette Rodrigo.

Contents

Tables and Figures

Foreword

The rapid economic growth of the People's Republic of China (PRC) during the last decades has led to very high CO_2 emissions and severe air pollution, prompting the national leadership to give greater strategic priority to CO_2 emissions reduction and air pollution alleviation. With its high CO_2 emissions, the Beijing–Tianjin–Hebei (BTH) region and surroundings have become a microcosm of the national challenge of sustaining rapid economic development while decisively curtailing the unabated growth of industrial emissions and waste. The development and implementation of well-coordinated climate, clean energy, and environmental policies in the BTH region is therefore imperative to effectively deal with this major problem over the long term.

At the request of the PRC, the Asian Development Bank (ADB) agreed to support the national government through a technical assistance (TA) project to help in formulating, validating, and monitoring a program to systematically cut down and control industrial emissions and waste. The project—under TA 9034, PRC: Developing Cost-Effective Policies and Investments to Achieve Climate and Air Quality Goals in the BTH Region— was jointly undertaken by the PRC's Ministry of Industry and Information Technology and ADB based on a memorandum of agreement they signed on 14 March 2016.

Beijing, Tianjin, and Hebei were selected as the focus municipalities and provinces for the study. During the implementation, however, Shandong was included in the study, with the province being a part of the greater BTH region and a source of high air pollution and high CO_2 emission levels owing to its large share of the PRC's heavy industry. In previous years, both the BTH region and Shandong set ambitious targets and exerted great efforts to mitigate their CO_2 emissions and air pollutants. But from now through 2035, they should fare much better in addressing the problem when the new integrated long-term, cost-effective strategy comes onstream.

The study has provided an integrated modeling framework to explore feasible pathways to reduce CO_2 emissions and air pollutants over the long term in the region. Several scenarios were simulated, and a package of policy measures were identified. The environmental goals attainment scenario represents one pathway to achieve the 2°C temperature control target and the National Ambient Air Quality Standards. A green-and-low-carbon transformation resulting from better energy efficiency and fuel switching combined with stringent end-of-pipe control measures for air pollutants is necessary. In particular, the study clearly shows how energy efficiency and fuel-switching measures can make crucial contributions to achieving these goals from now through 2035. Sectoral-level contributions in emissions mitigation have also been quantified. The energy sector would undergo a sustained transformation with increased use of renewable energy and imported electricity to substantially reduce the consumption of fossil fuel. The industrial sector would contribute most of the reductions in SO_2 emissions, primary fine particulate matter ($PM_{2.5}$) emissions, and CO_2 emissions, and the transport sector would contribute most of the reductions in NO_x emissions in 2035 relative to 2015 in the region. Understandings on major emissions reductions from sectors and activities would be of crucial value for decision-makers to prioritize policy interventions.

The results and conclusions of the study have been thoroughly reviewed by the PRC's policy makers and worldwide experts in each of the disciplines concerned. Their comments and inputs were fully considered in formulating the action plans and recommendations of the study.

In a statement delivered to the United Nations General Assembly on 22 September 2020, PRC President Xi Jinping announced that the country aims to achieve peak CO_2 emissions before 2030 and carbon neutrality before 2060. Now under preparation to attain these two goals are the corresponding Action Plans at the regional level.

ADB believes that in substantial measure, the insights and recommendations presented in this report form an integral part of the road map that the PRC's policy makers and experts are working on to fully achieve its goals for peak emissions and carbon neutrality.

M. Teresa Kho
Director General
East Asia Department
Asian Development Bank

Acknowledgments

This publication embodies the findings and recommendations of the study jointly undertaken by the Ministry of Industry and Information Technology (MIIT) of the People's Republic of China (PRC) and the Asian Development Bank (ADB) on how to decisively cut down and control the PRC's industrial emissions and waste. The study "Developing Cost Effective Policies and Investments to Achieve Climate and Air Quality Goals in the Beijing–Tianjin–Hebei Region" was led by Xuedu Lu, lead climate change specialist at ADB's East Asia Department (EARD), and supported by Gloria Gerilla-Teknomo, senior transport sector officer, EARD. Sujata Gupta, director, EARD, provided overall guidance and supervision. MIIT director Fengyuan Guo provided crucial support and guidance in preparing this publication. Dang M. Alcantara, project analyst, EARD, provided logistics support for the production of this publication.

The technical research was undertaken by a consultant team led by professor Xiliang Zhang of Tsinghua University, Beijing, with the following team members: Shuxiao Wang (Tsinghua University), Shiyan Chang (Tsinghua University), Jiayu Xu (Tsinghua University), Xi Yang (Tsinghua University), Yisheng Sun (Tsinghua University), Haotian Zheng (Tsinghua University), Lu Ren (Tsinghua University), Shengyue Li (Tsinghua University), Chengkui Gu (China Center for Information Industry Development), Jinsong Xiao (China Center for Information Industry Development), Molin Huo (State Grid), Lin Ma (China Center for Information Industry Development), Pengmei Li (China Center for Information Industry Development), Hongyu Zhang (Tsinghua University), Xiaodan Huang (Tsinghua University), Sining Ma (Tsinghua University), Siyue Guo (Tsinghua University), Da Zhang (Tsinghua University), Xunmin Ou (Tsinghua University), and Li Zhou (Tsinghua University). The draft of this report was reviewed by Valerie Karplus, assistant professor of the Massachusetts Institute of Technology (MIT), who provided very valuable inputs and suggested important revisions to the final manuscript.

Likewise, this publication greatly benefited from the comments and inputs of the following experts in their respective areas of specialization (sorted by surname): Bofeng Cai, Chinese Academy of Environmental Planning; Qinyang Dong, Beijing Green Industry Development Association; Hu Gao, Energy Research Institute, National Development and Reform Commission; Akiko Hagirawa, principal economist, ADB; Peng Jiang, Beijing E-Town International Investment & Development Co., Ltd; Yu Lei, Atmospheric Environment Institute, Chinese Academy for Environmental Planning in the Ministry of Ecology and Environment; Bing Li, China Metallurgical Industry Planning and Research Institute; Wenqiang Liu, China Center for Information Industry Development; Pradeep Perera, principal energy specialist, ADB; Qi Qiao, Chinese Research Academy of Environmental Science; Xionghui Qiu, North China Electric Power University; Zhuqiang Shao, China

Nonferrous Metals Industry Association; Duanhua Shi, China International Engineering Consulting Corporation; Xianchun Tan, Institute of Science and Technology Policy and Management, Chinese Academy of Sciences; Chunxiu Tian, Policy Research Center for Environment and Economy, Ministry of Ecology and Environment; Cindy Cisneros-Tiangco, principal energy specialist, ADB; Qing Tong, Tsinghua University; Liqiang Wang, Hebei Electronic Information Technology Institute; Xiaoping Xiong, Energy Research Institute, National Development and Reform Commission; Jingyu Yin, China Building Materials Federation; Yousheng Zhang, Energy Research Institute; National Development and Reform Commission; Yongping Zhai, senior advisor, Strategy Development Department, Tencent and former chief of Energy Group, ADB; and Hongchun Zhou, Development Research Center of the State Council.

Abbreviations

ADB	Asian Development Bank
APPCAP	Air Pollution Prevention and Control Action Plan
BTH	Beijing–Tianjin–Hebei
C-GEM	China-in-Global Energy Model
C-REM	China Regional Energy Model
CREM-HE	CREM-Health Effects Module
EGA	environmental goals attainment
ERSM	Extended response surface modeling
EOP	end-of-pipe
EOP-BAT	best available technologies measure package
EOP-CE	continued efforts measure package
ESP	electrostatic precipitator
ESP-FF	electrostatic-fabric integrated precipitator
FF	fabric filter
FGD	flue gas desulfurization
GDP	gross domestic product
HEFGD	high-efficiency flue gas desulfurization
LNB	low NO_x burning

MIIT	Ministry of Industry and Information Technology
NDC	Nationally Determined Contribution
NDRC	National Development and Reform Commission
NEA	National Energy Administration
NO_x	nitrogen oxide(s)
PM	particulate matter
$PM_{2.5}$	fine particulate matter
PRC	People's Republic of China
REACH	Regional Emissions Air-Quality Climate and Health
REPO	Renewable Electricity Planning and Operation Model
RSM	response surface modeling
S&H	services and household
SCC	social cost of carbon
SCR	selective catalytic reduction
SNCR	selective non-catalytic reduction
VSL	value of a statistical life
WESP	wet electrostatic precipitator
WHO	World Health Organization

Executive Summary

The combined area of Beijing–Tianjin–Hebei (BTH) and Shandong is a highly populated and rapidly growing industrial hub in the People's Republic of China (PRC). As this region has grown and urbanized, its air pollution and high carbon emissions have continuously grown to levels that pose a major constraint to the region's development, ecology, and public well-being.

A set of measures for addressing air pollution and CO_2 emissions has been introduced in the PRC in recent years; consequently, the annual average $PM_{2.5}$ concentrations in Beijing, Tianjin, Hebei, and Shandong declined between 2013 and 2020. In Beijing, these concentrations declined from 89.5 $\mu g/m^3$ in 2013 to 38 $\mu g/m^3$ in 2020; in Tianjin they declined from 96 to 48 $\mu g/m^3$; in Hebei from 108 to 44.8 $\mu g/m^3$; and in Shandong from 98 to 46 $\mu g/m^3$. Still, these are much higher than the 35 $\mu g/m^3$ level stipulated by the PRC's National Ambient Air Quality Standards, as well as the 5 $\mu g/m^3$ level stipulated by World Health Organization (WHO) guidelines.

There is likewise a need to significantly cut down the region's CO_2 emissions to support the PRC's contribution toward achieving the 2°C temperature control limit on postindustrial temperature rise as set forth in the 2015 Paris Climate Accord. Substantial progress toward these goals will require enhanced and more ambitious policy objectives and sustained support.

Launched under the cooperation framework provided by the Ministry of Industry and Information Technology and the Asian Development Bank, the study has been carried out by Tsinghua University working in close collaboration with the China Center for Information Industry Development. The study has three major objectives:

(i)   Develop climate and air quality goals for BTH and Shandong up to the year 2035.
(ii) Identify policies and programs to achieve these goals.
(iii) Provide an integrated modeling framework for evaluating the effects of these policies and programs on the national economy, the environment, and human health and well-being.

Methodology

The policy scenarios in the study were simulated using the Regional Emissions Air-Quality Climate and Health (REACH) framework. In the research study, the REACH framework integrated the energy-economic model of the China Regional Energy Model with the Chinese Emission Inventory and air-quality simulation model called the Extended Response Surface Model. This publication focuses on two scenarios: the Current Policies and the environmental goals attainment (EGA). More information can be found in Appendix 1.

The Current Policies scenario considers the impact of implementing existing policies and plans that various signatory governments had already announced they would undertake under the 2015 Paris Agreement Climate Accord. The EGA scenario simulates policies that would limit $PM_{2.5}$ concentrations to no higher than 35 µg/m³ by 2035 for the four municipalities and provinces, and that would reduce CO_2 emissions in line with the agreed upon 2°C maximum climate change target.

The study analyzed the benefits and costs of measures in the EGA scenario relative to the Current Policies scenario. The contributions of individual EGA scenario measures toward achieving both the air quality and climate goals were quantified. Scenario comparisons were then undertaken to form the basis for the policy recommendations of the study.

Key Findings

1. **The EGA scenario represents one feasible pathway to achieve CO_2 mitigation and air-quality improvement goals in BTH and Shandong that would not impede nor markedly diminish economic growth.**

   Compared to the Current Policies scenario, the EGA scenario would achieve both air quality and climate change targets for the region without unduly curtailing economic activity. Relative to their 2015 levels, the Current Policies scenario would reduce the annual mean $PM_{2.5}$ concentration in 2035 by 55% to 36 µg/m³ in Beijing, by 36% to 45 µg/m³ in Tianjin, by 50% to 38 µg/m³ in Hebei, and by 47% to 41 µg/m³ in Shandong.

   With additional measures in the EGA scenario, the annual mean $PM_{2.5}$ concentration would decline by 71% to 24 µg/m³ in Beijing, by 51% to 34 µg/m³ in Tianjin, by 62% to 29 µg/m³ in Hebei, and by 59% to 31 µg/m³ in Shandong. These reductions would achieve the regulated levels that the PRC has stipulated.

   The region's CO_2 emissions totaled 1.83 billion tons in 2015. In the Current Policies scenario, CO_2 emissions would peak by 2025 at 1.94 billion tons, while under the EGA scenario, CO_2 emissions would decrease to 1.72 billion tons by 2025, and further to 1.25 billion tons by 2035.

2. **In the EGA scenario, the energy system in the region would undergo sustained transformation from now until 2035.**

   Even after the recent implementation of new controls and emissions standards, reliance on coal as fuel for industrial production and for heating would remain as the main cause of severe air pollution and CO_2 emissions in the region. Decarbonization needs to be pursued to reduce both impacts.

In 2015, coal consumption, including net imports of coke, totaled about 8 Mtce in Beijing, 41 Mtce in Tianjin, 236 Mtce in Hebei, and 290 Mtce in Shandong, representing roughly 0.3%, 1.5%, 8.5%, and 10.5% of the national total, respectively.

The EGA scenario approach showed that, under a certain set of assumptions, the National Ambient Air Quality Standard of 35 $\mu g/m^3$ could be achieved in all four municipalities and provinces by 2035. The first step would be limiting coal consumption by 2025 to no more than 0.7 Mtce in Beijing, 30 Mtce in Tianjin, 215 Mtce in Hebei, and 272 Mtce in Shandong, and then by 2035 to no more than 0.5 Mtce in Beijing, 14 Mtce in Tianjin, 146 Mtce in Hebei, and 181 Mtce in Shandong, respectively.

Increased utilization of renewable energy would substantially reduce consumption of fossil fuel—especially coal—in the electricity sector. In the 2015 electricity supply, the share of renewable energy (wind, solar, hydro, and biomass) was about 4% in Beijing, 2% in Tianjin, 9% in Hebei, and 5% in Shandong. These shares were much lower than the national average share of 24% in that same year.

One way to significantly increase this renewable energy contribution would be by increasing imports of electricity that is being generated by other provinces outside the region. A second way would be to substantially increase the region's own installed renewable energy capacity.

Relative to 2015, the EGA scenario projections assumed that the installed capacity of wind power by 2035 would have increased from 200 MW to 3.5 GW in Beijing, from 320 MW to 2.5 GW in Tianjin, from 10 GW in 2015 to 96.5 GW in Hebei, and from 7.2 GW to 47 GW in Shandong.

Solar power is projected to increase from 165 MW to 12.5 GW in Beijing, from 125 MW to 5.9 GW in Tianjin, from 2.8 GW to 79 GW in Hebei, and from 1.3 GW to 41 GW in Shandong.

Imported electricity in the four subregions is also projected to increase to total annual imports of 480 TWh/year by 2035, a level that would account for about 27% of the total electricity demand of the entire region.

3. **In the EGA scenario, the industrial sector would contribute around 58% of the reduction in SO_2, 33% of the reduction in NO_x, 55% of the reduction in primary $PM_{2.5}$, and 54% of the reduction in CO_2 in 2035 relative to 2015.**

Under the EGA scenario, reductions in emissions of SO_2, NO_x, and primary $PM_{2.5}$ from industrial facilities would contribute to a very substantial improvement in air quality.

The REACH model projects that, due to the combined impact of policies that enhance energy efficiency, promote fuel switching, and strengthen end-of-pipe air pollutants control in 2035 relative to 2015, the SO_2 emissions of the industrial sector would fall by 86% in Beijing, by 79% in Tianjin, by 82% in Hebei, and by 87% in Shandong.

NO_x emissions would be lowered by 73% in Beijing, by 61% in Tianjin, by 71% in Hebei, and by 59% in Shandong.

Primary $PM_{2.5}$ would go down by 5% in Beijing, by 69% in Tianjin, by 71% in Hebei, and by 79% in Shandong.

During the same period, the EGA scenario policies would bring about a decline in industrial CO_2 emissions of 30% in Beijing, 39% in Tianjin, 39% in Hebei, and 39% in Shandong.

4. The EGA scenario would require integrated planning efforts and additional policy interventions to achieve the climate and air quality goals in tandem.

In the EGA scenario, a green-and-low-carbon transformation would result from better energy efficiency and fuel switching combined with stringent end-of-pipe control measures for air pollutants. Compared to the Current Policies scenario, energy efficiency and fuel-switching measures across provinces in 2035 would contribute 60%–72% of the $PM_{2.5}$ concentration reduction, while end-of-pipe control measures would at the same time deliver 28%–40% of the total reductions. This shows how important the role of energy efficiency and fuel-switching measures would be in improving air quality in the four municipalities and provinces.

In all subregions, energy efficiency and fuel switching incentivized by a CO_2 price would contribute the largest share of reductions. To build on the substantial progress achieved in recent years, additional end-of-pipe control measures would need to be systematically implemented and the installation and operation of control equipment required on a mandatory basis. In the EGA scenario, these two measures would be crucial to the attainment of the National Ambient Air Quality Standard by 2035.

5. To implement the EGA scenario, caps for CO_2 emissions and $PM_{2.5}$ concentrations that have been translated into binding targets, together with supplementary measures, are recommended to mobilize the necessary attention and investments.

Regionwide, binding targets are projected to mobilize the necessary amount of investment into energy-efficiency improvements and low-carbon fuel switching. Once these broader goals are in place, renewable portfolio standards and a green electricity dispatching scheme would play a larger role in raising the share of renewable energies in the electricity supply of municipalities and provinces.

In parallel, measures would be needed to accommodate the intermittency of the renewable energy supply as well as to increase the stability of the electricity system. Regulations that ban investments in production capacity expansion would likewise be needed for high-polluting sectors, coupled with incentives to substitute cleaner technologies.

Finally, the stringency of performance standards for SO_2, NO_x, and particulate matter emissions would need to be raised in all highly polluting industrial sectors.

6. **Substantial incremental investments would be needed to achieve a radical green-and-low-carbon transformation in the region. These investments should focus on the service and non-energy-intensive industries.**

In the EGA scenario, the model had simulated an additional $573 billion in total incremental capital investment between 2020 and 2035 for energy-efficiency improvement and low-carbon fuel switching in the various industrial regions. A total of $118 billion would be accounted for by Beijing, $77 billion by Tianjin, $125 billion by Hebei, and $253 billion by Shandong.

Fifty-eight percent or most of the investments would flow to the services sector, and 25% to the other non-energy-intensive industries. Outside of the fossil energy-intensive sectors, a sizable share of investment in the EGA scenario should flow to public transport, 8%; to the power sector, 2%; and to the chemicals sector (including refined oil), 6%.

1 Introduction

The combined area of Beijing–Tianjin–Hebei (BTH) and Shandong in the People's Republic of China (PRC) is highly industrialized and densely populated and is currently undergoing rapid economic growth and urbanization. The high fossil fuel intensity of economic activity underpins the region's status as a hot spot in the PRC's air pollution situation and as a major source of the climate-warming greenhouse gas CO_2. As a result, the region has long figured centrally in national efforts to improve air quality and mitigate climate change.

Ongoing control efforts in the region have improved air quality substantially, but there is still a long way to go. When the PRC's Air Pollution Prevention and Control Action Plan (APPCAP) was rolled out in 2013, the annual average fine particulate matter ($PM_{2.5}$) concentration was $90\,\mu g/m^3$ in Beijing, $96\,\mu g/m^3$ in Tianjin, $108\,\mu g/m^3$ in Hebei, and $98\,\mu g/m^3$ in Shandong. Since then, many measures, e.g., strict end-of-pipe emissions control, have been deployed to address the air pollution problem.

These measures have succeeded in improving air quality. The annual average $PM_{2.5}$ concentrations fell to between $38\,\mu g/m^3$ and $48\,\mu g/m^3$ in the four subregions in 2020. However, the pollution levels remained much higher than $35\,\mu g/m^3$, which is the target level stipulated by the PRC's National Ambient Air Quality Standards. They are likewise higher than the $5\,\mu g/m^3$ level set by the air quality guidelines of the World Health Organization (WHO). The region also needs to further reduce CO_2 emissions to be able to support the PRC's contribution to the global 2°C target set forth in the Paris Climate Agreement.

Integrated solutions are needed to address the region's air pollution and climate change because the same underlying fossil emissions sources contribute to both problems. The few studies that have addressed these problems in BTH had focused separately on air pollution[1] and CO_2 emissions.[2] The studies tended to concentrate on emissions rather than on the ambient air quality or climate change targets, particularly the thresholds recommended in the National Ambient Air Quality Standards or the 2°C temperature control target.

---

[1] J. Hao. 2018. *Strategy on Regional Economic Development and Joint Air Pollution Control in Beijing–Tianjin–Hebei Region*. Beijing: Science Press; W. Wu, et al. 2017. Assessment of $PM_{2.5}$ Pollution Mitigation due to Emission Reduction from Main Emission Sources in the Beijing–Tianjin–Hebei Region. *Environmental Science*. 38(3), pp. 867–75; N. Li, et al. 2019. Does China's Air Pollution Abatement Policy Matter? An Assessment of the Beijing–Tianjin–Hebei Region Based on a Multi-Regional CGE Model. *Energy Policy*, 127, pp. 213–27.

[2] Q. Yan, et al. 2019. Coordinated Development of Thermal Power Generation in Beijing–Tianjin–Hebei Region: Evidence from Decomposition and Scenario Analysis for Carbon Dioxide Emission. *Journal of Cleaner Production*. 232, pp. 1402–17; J. Zhou, et al. 2018. Scenario Analysis of Carbon Emissions of Beijing–Tianjin–Hebei. Energies, 11 (14896); C. Wang, et al. 2019. Structural Decomposition Analysis of Carbon Emissions from Residential Consumption in the Beijing–Tianjin–Hebei Region, China. *Journal of Cleaner Production*, 208, pp. 1357–64.

Moreover, the cost and effectiveness of feasible integrated packages of policy measures that would address climate change and air pollution have not previously been evaluated in a context that would account for cross-sectoral linkages and feedback.

The new study addresses these gaps, with the focus centering on the following research questions:

(i)   In which sectors can technology deployment strategies help to simultaneously address both CO_2 emissions and air quality?

(ii)   What feasible measures to incentivize efficiency and fuel switching, and implementing what end-of-pipe control strategies, would enable the attainment of the country's air quality target and climate goals?

(iii) What are the costs and benefits of using specific measures to address CO_2 emissions and air quality goals?

(iv)   What incremental capital investment for implementing the projected policies would achieve both the air quality and climate change goals for the region?

The remainder of this report has been organized as follows:

Chapter 2 describes the contribution of the region's economic activities to the state of air quality and climate in the regions as well as the challenges to addressing both.

Chapter 3 reviews the measures and policies previously adopted to address climate change and air pollution in the regions.

Chapter 4 introduces the integrated modeling framework applied in the study.

Chapter 5 describes the application of an integrated modeling framework to analyze alternative policy scenarios.

Chapter 6 concludes the report with the insights and policy recommendations based on the analysis of the various scenarios.

The appendixes provide additional details on the design of the scenarios.

2 Regional Background

The Beijing–Tianjin–Hebei (BTH) region and Shandong constitute a major center of economic, social, and cultural activities in the People's Republic of China (PRC). Together, the two municipalities (Beijing, Tianjin) and two provinces (Hebei, Shandong) accounted in 2020 for approximately 15% of the PRC's population and 16% of the country's gross domestic product (GDP).

The area has been designated one of the most severely polluted regions in the PRC, with a particularly high incidence of heavy air pollution. In 2013, the average annual $PM_{2.5}$[3] concentration in every subregion was roughly equal to or greater than 90 μg/m³ (Figure 1).

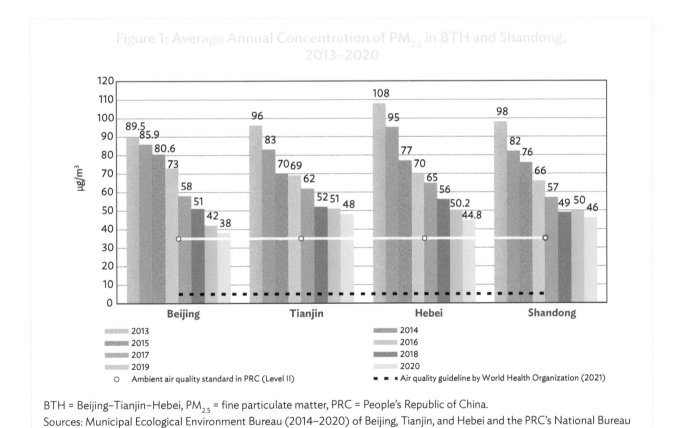

Figure 1: Average Annual Concentration of $PM_{2.5}$ in BTH and Shandong, 2013–2020

BTH = Beijing–Tianjin–Hebei, $PM_{2.5}$ = fine particulate matter, PRC = People's Republic of China.
Sources: Municipal Ecological Environment Bureau (2014–2020) of Beijing, Tianjin, and Hebei and the PRC's National Bureau of Statistics (2014–2020).

---

3    $PM_{2.5}$ is particulate matter with a mean aerodynamic diameter of 2.5 μm or less.

In recent years, air quality had improved significantly due to the introduction of a set of air pollution mitigation measures. By 2020, the average annual $PM_{2.5}$ had decreased to about 38 µg/m³ in Beijing, 48 µg/m³ in Tianjin, 44.8 µg/m³ in Hebei, and 46 µg/m³ in Shandong.

However, the region remains among the most polluted areas in the PRC. According to the air quality comprehensive index of the PRC's 168 cities that was issued by the Ministry of Ecology and Environment in 2020, five of the 10 cities with the worst air quality[4] are located in BTH and Shandong.

Emissions of the major air pollutants SO_2, NO_X, and particulate matter (PM) are responsible for the region's high $PM_{2.5}$ concentrations. Total emissions are shown in Figure 2.

In 2015, the industrial sector accounted for 54%, 38%, and 54% of the region's SO_2, NO_X, and primary $PM_{2.5}$ emissions, respectively. For the power sector, these shares were 26% for SO_2, 25% for NO_X, and 7% for $PM_{2.5}$.

The transport sector accounted for 3% for SO_2, 31% for NO_X, and 2% for $PM_{2.5}$, while the services and household (S&H) sector accounted for 17% for SO_2, 4% for NO_X, and for 23% $PM_{2.5}$ (Figure 2).

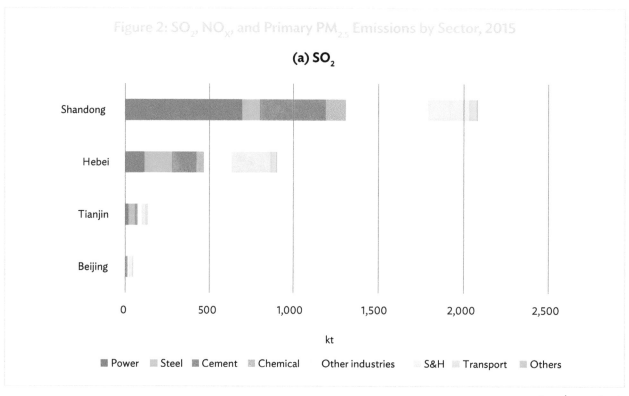

Figure 2: SO_2, NO_X, and Primary $PM_{2.5}$ Emissions by Sector, 2015

(a) SO_2

continued on next page

---

4   Shijiazhuang (Hebei Province), Tangshan (Hebei Province), Handan (Hebei Province), Zibo (Shandong Province), and Xingtai (Hebei Province).

Figure 2 *continued*

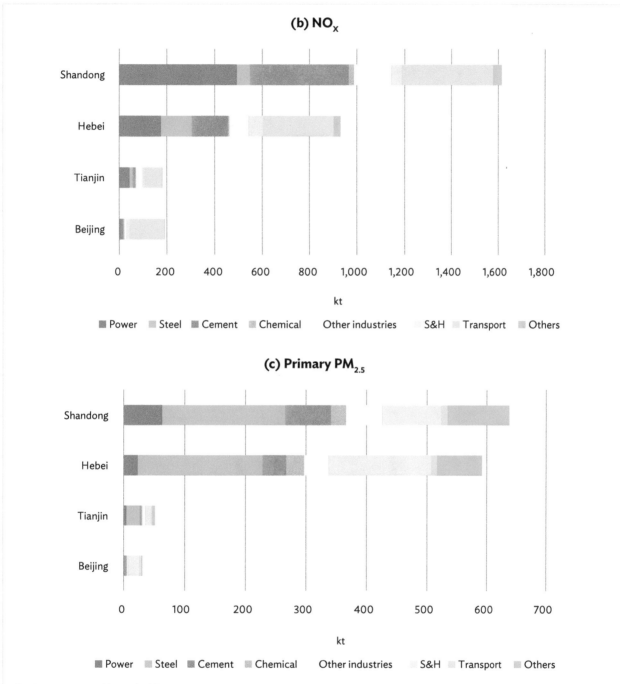

S&H = services and household.

Notes: (i) Steel sector includes iron and steel and nonferrous industries. (ii) The cement sector refers to the nonmetallic industries. (iii) The chemical sector includes chemistry and refined oil industries. (iv) The Others sector includes mining, construction, transportation equipment manufacturing, food, textiles, electronic equipment manufacturing, machinery manufacturing, water, and several sectors with more limited contributions to economic output.

Source: Chinese emissions inventory.

Emissions in Hebei and Shandong were substantially greater than those of either Beijing or Tianjin. Due to differences in their industrial composition, activity levels, and fuel mix, the contribution of each subregion to total regional emissions varies by type of pollutant (Figure 3).

Shandong is the biggest emitter of SO_2, NO_x, and primary $PM_{2.5}$, contributing 66%, 55%, and 49%, respectively, to the total of each pollutant in the regions. In 2015, Hebei Province contributed 29% of SO_2, 32% of NO_x, and 45% of primary $PM_{2.5}$ emissions.

The contributions of Beijing and Tianjin to total emissions of each air pollutant were much lower compared to those of the two provinces.

BTH and Shandong CO_2 emissions totaled 1.8 billion tons in 2015,[5] or approximately 45% compared to Europe's total CO_2 emissions.[6] Of this total, the power sector accounted for 43%, the industrial sector 44%, the transportation sector 6%, and the S&H sector 6%. Breakdowns by subregion are shown in Figure 4.

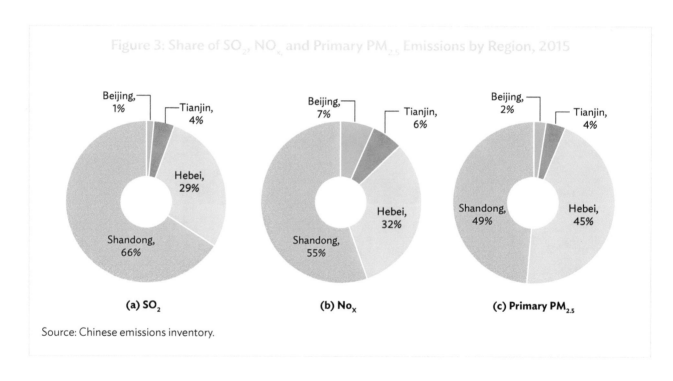

Figure 3: Share of SO_2, NO_x and Primary $PM_{2.5}$ Emissions by Region, 2015

(a) SO₂   (b) Noₓ   (c) Primary PM₂.₅

Source: Chinese emissions inventory.

---

5   CO_2 emissions in the report refer to those that are domestic energy-related, and that are defined as emissions from the use of fossil fuels (coal, liquid fuels, and natural gas). Emissions and removals from land-use change and forestry are not included.

6   The total CO_2 emission of Europe was around 4 billion tons in 2015. Also see BP. 2018. BP Statistical Review of World Energy 2018. https://www.bp.com/en/global/corporate/energy-economics/statistical-review-of-world-energy.html.

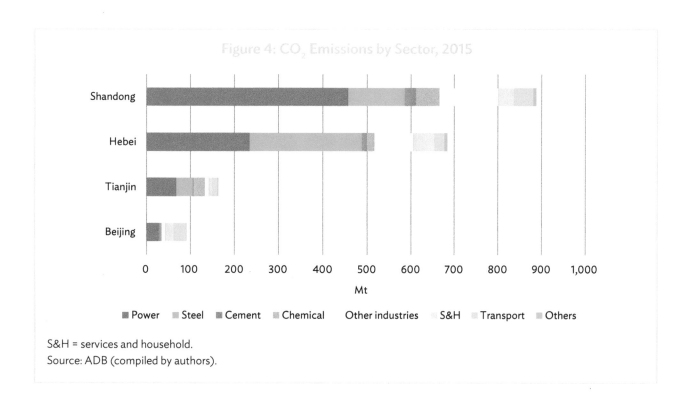

Figure 4: CO_2 Emissions by Sector, 2015

S&H = services and household.
Source: ADB (compiled by authors).

Table 1: Population, GDP, and Industrial Added Value in BTH and Shandong, 2020

Region	Population/ 10,000	GDP/ CNY100 million	Value Added of Secondary Industry/ CNY100 million	Secondary Industrial Added Value in GDP (%)	GDP per Capita/ CNY	Total Energy Consumption/ 10,000 tce (2019)
Beijing	2,189	36,103	5,716	16	164,927	7,360
Tianjin	1,387	14,084	4,804	34	101,541	8,241
Hebei	7,461	36,207	13,597	38	48,528	32,545
Shandong	10,153	73,129	28,612	39	72,027	41,390
Total	21,190	159,522	52,730	33	75,282	89,536

BTH = Beijing–Tianjin–Hebei, GDP = gross domestic product.
Source: National Bureau of Statistics Yearbook. 2020. China Annual Statistical Yearbook.

Table 1 summarizes the differences in population, economic composition, industrial structure, and energy mix. Shandong has the second-largest population of any province or autonomous region in the PRC, with a total population in 2019 of about 100 million. The municipalities of Beijing and Tianjin have a population of about 20 million and 15 million, respectively. Beijing is the wealthiest, with a GDP per capita income of more than CNY160,000, while Hebei's was around CNY46,000, and Shandong's, about CNY70,600.

The major causes of severely degraded air quality in the region were its high industry share in economic activity and its continuing reliance on coal as fuel for industrial production and heating.

In 2019, secondary industry contributed 34% of the region's economic value-added, with steel, coke, and flat glass accounting for 33%, 21%, and 28%, respectively, of total nationwide production (Figure 5).

At present, the pollution and CO_2 intensity of the region's economy are not compliant with the PRC's goals of air-quality improvement and deep decarbonization. As mentioned, current $PM_{2.5}$ concentrations exceed the guidance of national and WHO standards. The PRC has pledged to address climate change under the Paris Agreement by achieving peak CO_2 emissions no later than 2030 and reducing CO_2 emissions per unit of GDP by 60%–65% by 2030 relative to 2005 levels.

If CO_2 emissions continue to rise as the region's economy grows, achieving national climate goals would become more difficult. Therefore, transformation of industry should be viewed as a crucial opportunity to support economic and energy pathways that could address both goals.

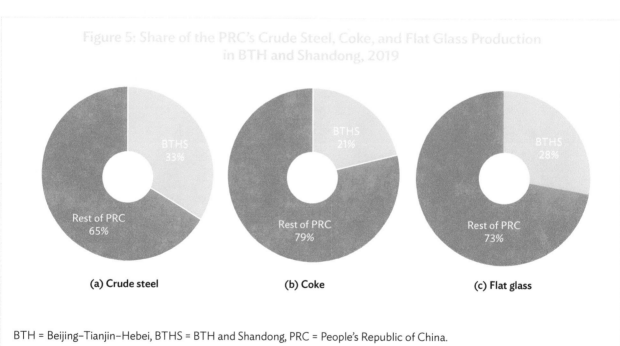

Figure 5: Share of the PRC's Crude Steel, Coke, and Flat Glass Production in BTH and Shandong, 2019

(a) Crude steel   (b) Coke   (c) Flat glass

BTH = Beijing–Tianjin–Hebei, BTHS = BTH and Shandong, PRC = People's Republic of China.
Source: National Bureau of Statistics Yearbook. 2020. China Annual Statistical Yearbook.

3 Existing Regional Policies and Measures to Address Climate Change and Air Pollution

Over the past decade, climate change has become a very important policy priority in the PRC. The PRC's Nationally Determined Contribution (NDC) in 2015 included these three major energy-related objectives:

(i)   to enact policies that would achieve peak CO_2 emissions around 2030 and to make best efforts to achieve this peak even earlier;

(ii) to reduce the CO_2 intensity of the PRC's economy by 60%–65% by the year 2030 relative to its 2005 level; and

(iii) to increase the nonfossil share of primary energy consumption by 2030 to around 20%.

Climate Change Targets

To realize these objectives, the PRC has introduced targets for energy intensity, CO_2 intensity, and nonfossil fuel share into its national Five-Year Plans for economic and social development.

The targets in the PRC's Thirteenth Five-Year Plan (2016–2020) were as follows:

(i)   to reduce the energy intensity of the economy by 15% relative to 2015;

(ii) to reduce the CO_2 intensity of the economy by 18% over the same period; and

(iii) to increase the share of nonfossil fuels in the primary energy supply to 15% by 2020.

Efforts were taken to meet these target levels. The CO_2 intensity decreased by 18.8% during the period and the share of nonfossil fuels in the primary energy supply increased to 15.9% by 2020.

To distribute the responsibility for fulfilling national climate change mitigation obligations, the targets for national energy and CO_2 intensity reduction have been disaggregated to subordinate levels of government. Based on each province's stage of development and resource endowments, the provincial CO_2 intensity targets under the Thirteenth Five-Year Plan are divided into five categories.[7] Beijing, Tianjin, Hebei, and Shandong must attain a CO_2 intensity reduction target of 20.5% by 2020, relative to 2015 (Table 2).

---

[7]   In the national Thirteenth Five-Year Work Plan on Greenhouse Gas Control, the CO_2 intensity reduction targets of 30 provinces are divided into five categories. The plan requires a 20.5% reduction in CO_2 intensity in Beijing, Tianjin, Hebei, and Shandong—along with Shanghai, Jiangsu, Zhejiang, and Guangdong—in 2020 relative to 2015, which is more stringent than other provinces. The CO_2 intensity reductions required in other provinces range from 12% to 19.5%.

Table 2: Climate and Air Quality Targets, 2020

Region	CO_2 Intensity Reduction Target by 2020 Relative to 2015 (%)	Target for Annual Average $PM_{2.5}$ Concentration by 2020 (µg/m³)
PRC	18[a]	None
Beijing	20.5[a]	56[b]
Tianjin	20.5[a]	52[c]
Hebei	20.5[a]	55[d]
Shandong	20.5[a]	49[e]

PRC = People's Republic of China.
Sources: (a) The PRC's Thirteenth Five-Year Plan Program for Controlling Greenhouse Gas Emissions; (b) Thirteenth Five-Year Plan for Environmental Protection and Ecological Construction for Beijing (2016); (c) Three Year Plan to Win the Blue Sky Defense War of Tianjin (2018); (d) Three Year Action Plan to Win the Blue Sky Defense War for Hebei (2018); (e) Action Plan to Win the Blue Sky War for Shandong (2018).

Beijing and Tianjin have in addition formulated their own mitigation targets to control their absolute emissions. Beijing targeted a CO_2 emissions peak at the latest by 2020, while Tianjin aims to reach its peak CO_2 emissions by around 2025. The governments of Beijing, Tianjin, Hebei, and Shandong have further disaggregated to their constituent cities and districts their targets for energy and CO_2 intensity reduction.

On 22 September 2020, President Xi Jinping of the PRC, addressing the General Debate of the 75th Session of the United Nations General Assembly, declared that the PRC would scale up its Intended Nationally Determined Contributions by adopting more vigorous policies and measures to strive for CO_2 emissions peak before 2030, and that the PRC intends to achieve carbon neutrality by 2060. The corresponding Action Plan on Achieving CO_2 Emission Peak by 2030 has been released and the local targets to reduce CO_2 emissions in the Fourteenth Five-Year Plan (2021–2025) are in the formulation.

Stringent Targets to Improve Air Quality

Air pollution is a major concern in the PRC, occupying with each passing decade an increasingly prominent place in the country's policy agenda. For this reason, Air Pollution Prevention and Control Action Plan (APPCAP) was introduced by the State Council in 2013. At that time, it constituted the PRC's most stringent suite of air pollution control measures.

The APPCAP included binding ambient air-quality improvement targets over the period 2012–2017, with the BTH region—along with the Pearl River Delta and Yangtze River Delta regions—mandated to pursue more stringent targets. In BTH, the APPCAP targeted a 25% reduction by 2017 in their annual average $PM_{2.5}$ concentrations compared to 2012. In addition, the Action Plan set an annual concentration target for Beijing by 2017 of approximately 60 µg/m³.

Under the Three-Year Action Plan for Winning the Blue Sky Defense Battle (2018–2020) that had since superseded the APPCAP, each province updated their air quality targets through 2020 in either their Thirteenth Five-Year Plans or air quality programs. In their plans, Beijing, Tianjin, Hebei, and Shandong pledged to decrease the $PM_{2.5}$ concentration by 2020 to 56 µg/m³, 52 µg/m³, 55 µg/m³, and 49 µg/m³, respectively. These have been achieved by the four subregions (see Figure 1).

Medium- and long-term air-quality attainment plans for the region are currently being drafted.

Existing Policies and Measures for Industry

The industrial sectors of Beijing, Tianjin, Hebei, and Shandong are major contributors to the regional economy, together accounting in 2019 for about 28% of total economic activity. Over the past decade, however, overcapacity in many industries and concerns about air pollution prompted the government to

(i)   encourage restructuring of the regional economy;

(ii)   target improvements in efficiency; and

(iii) incentivize fuel switching, especially from coal to natural gas and renewable energies.

Other policies had focused on cleaning up air pollutant emissions by requiring facilities to adopt end-of-pipe control measures.[8] While these levers have supported progress on air quality and climate goals to date, the assessment of the study still found considerable additional potential to use them to achieve tougher targets on the horizon.

Industrial Restructuring

The industrial restructuring policies aim to change the structure of the industrial sector by eliminating excess capacity, fostering new strategic industries, and implementing off-peak production. While they do not explicitly target reductions in air pollution and CO_2 emissions, pursuing these policies more aggressively would affect these emissions by significantly altering the composition of the regional economy.

---

[8]   End-of-pipe control measures refer to methods or technologies used to remove already formed contaminants from a stream of air, water, waste, or other product. These techniques are called "end-of-pipe" as they are normally implemented as a last stage of a process before the stream is disposed of or delivered such as scrubbers on smokestacks and catalytic converters on automobile tailpipes that reduce emissions of pollutants after they have formed. Also see United Nations Environment Programme. 2019. *Air Pollution in Asia and the Pacific: Science-based Solutions.* https://www.ccacoalition.org/en/resources/air-pollution-asia-and-pacific-science-based-solutions-summary-full-report.

Several policies have already been implemented to achieve industrial restructuring in BTH and Shandong, as follows:

Mandated Capacity Reductions in Targeted Energy-Intensive industrial Sectors

Following the State Council guidance released in 2013, the national government had introduced policies to reduce overcapacity in energy-intensive industries. Municipal and provincial governments had responded with detailed implementation plans.

In Beijing, most of the energy-intensive industries had been phased out prior to 2013, and a stringent production capacity cap to 4 million tons by 2017 had been set for the cement industry.

Tianjin limited its production capacity for iron and steel in the municipal administrative area by 2017 to 20 million tons or less, and of its cement (clinker) production capacity to 5 million tons or less.

Hebei introduced caps on major energy-intensive production capacity in 2014. It required reductions from 2014 to 2017 of 60 million tons in iron and steel, of 60 million tons in cement, and of 30 million weight boxes of flat glass.

Shandong's plan required cumulative reductions from 2016 to 2020 of 9.7 million tons of iron production capacity and 15 million tons of crude steel production capacity. Targets for the region during this period were met or even surpassed.

In Beijing, the cement industry capacity decreased from 10 million tons in 2011 to less than 4 million tons by 2017.

Tianjin reduced its crude steel production capacity by 7.5 million tons from 2013 to 2017, and achieved its target to limit iron and steel production capacity within 20 million tons inside municipal areas.

Hebei overfulfilled its targets. From 2013 to 2017, the cumulative reductions in its steelmaking capacity, iron production capacity, cement production capacity, and flat glass production capacity totaled 70 million tons, 64 million tons, 71 million tons, and 72 million weight boxes, respectively.

Additional capacity reduction targets had been set for 2020. To meet them, 1,200 energy-intensive, highly polluting firms in Beijing have already been phased out to date.

Tianjin had targeted reductions in crude steel production capacity by 6.9 million tons from 2018 to 2020.

Hebei had set new production capacity caps by 2020 of 200 million tons for iron and steel, 200 million tons for cement, and 200 million weight boxes of flat glass.

Shandong had planned to limit by 2020 its annual capacity of cement clinker and flat glass to about 80 million tons and 70 million weight boxes, respectively.

The capacity cuts had typically been pursued or achieved through a combination of the following measures:

(i)   **Regulating or banning of investments.** Fixed-asset investments are strictly regulated for the energy-intensive industries, and bans are imposed on new construction.

(ii)   **Forcing retirement of facilities.** Facilities are ordered to shut down permanently if they do not meet energy efficiency standards, or to surrender their pollutant emissions permits if their emissions equal or exceed the permissible levels.

(iii) **Offering financial incentives.** The central government had allocated CNY100 billion to subsidize or reward firms that cut their excess production capacity or adjust product portfolios. To be eligible for incentives, municipal and provincial governments have been mandated to partially match these funds.

Fostering Strategic New Industries

Part of the economic restructuring would involve promoting the development of strategic new industries in the region.

In Beijing, only firms manufacturing products classified as having high value-added with low energy consumption and low pollution are allowed to enter the market.

Hebei had identified seven key strategic industries in its Thirteenth Five-Year Plan for Industrial Transformation and Upgrading. These include advanced equipment manufacturing, next-generation information technology, lifestyle and health, new energy, new materials, energy-saving and environmental protection, and specialty robotics.

In Shandong, the Action Plan for Promoting Industrial Transformation and Upgrading (2015–2020) had provided a list of key areas for upgrading, including four new industries, namely high-end equipment manufacturing, modern medicine, next-generation information technology, and new materials. Preferential policies would foster the strategic new industries and provide access to land and tax incentives.

Implementing Off-Peak Production

To address severe air pollution in autumn and winter, environmental agencies have been mandated to order the stoppage of industrial production in certain areas. Subjected to off-peak production stoppage are these industries: iron and steel, coking, casting, cement, flat glass, nonferrous metal, and refining/chemicals.

Other industries may be added at the discretion of the provincial or municipal government.

Increasing Energy Efficiency

Improvements in industrial energy efficiency are currently being targeted by the following multiple national programs in the PRC:

(i)   the Thirteenth Five-Year Comprehensive Work Plan for Energy Conservation and Emissions Reduction that has been released by the State Council;

(ii)   the Thirteenth Five-Year Action Plan for National Energy Conservation that was jointly issued by 13 ministry-level entities, among them the National Development and Reform Commission (NDRC), the Ministry of Science and Technology, the Ministry of Industry and Information Technology (MIIT), the Ministry of Finance (MOF), etc.; and

(iii) the Industrial Green Development Plan (2016–2020) that is being implemented by the MIIT.

Accelerating Deployment of Advanced Energy-Saving Technologies

Government programs would strongly focus on helping firms undertake energy-saving measures by providing them with information and incentives. The approved energy-saving technologies and products were identified in the List of Recommended Industrial Energy-Saving Technology and Equipment and in the List of the Recommended Advanced and Applicable Technologies.

As of 2013, direct financial support for the adoption of energy-efficiency measures in firms had been phased out and replaced by new programs targeted for funding. The MOF and MIIT have jointly launched a program on Green Manufacturing Integration Systems.

So far, more than 360 projects have received funding from the new program, which focuses on non-energy-intensive manufacturing including machinery, electronics, food, textiles, home appliances, and large-scale equipment.

Energy Efficiency Leadership Program for Key Energy-Intensive Industries

The Energy Efficiency Leadership Program is a ministerial-level initiative that has been underway since its launching in 2014. To qualify for this program, the energy-intensive firms and public institutions should meet or exceed specific energy-efficiency benchmarks. For example, the energy consumption per unit product of major industrial processes in an energy-intensive firm needs to meet or exceed the advanced value in the Norm of Energy Consumption per Unit Product, and achieve the industry-leading level.

Industries covered by this program include thermal power, crude oil processing, iron and steel, ethylene, synthetic ammonia, electrolytic aluminum, flat glass, and cement. The MIIT, NDRC, and National Energy Administration (NEA) are leading this program and they are mandated to identify high-performing firms.

This program is designed to raise energy efficiency through two channels. The first is to encourage leadership by example. The second is to validate recommended energy-efficiency interventions so that the government could justify moving from voluntary to mandatory standards.

Municipal and provincial governments have been given the latitude to adjust efforts to promote energy-efficiency consonant with local conditions. For example, the Beijing government had given a financial reward to companies and institutions that have won the title of "Beijing Energy Efficiency Leader."

In Beijing, the energy-efficiency leaders designated by the national program are being given special treatment in the approval of fixed-assets investment projects and increased eligibility to receive governmental financial support for their investment projects.

Fuel Switching and Renewable Energy Development

Policies have been instituted to encourage fuel switching, particularly in increasing share of renewable energy and in better addressing both air quality and climate change goals.

Small Coal Boilers

To promote the phaseout of small-scale coal boilers, various specific policy measures have been adopted by provincial and municipal governments. Very stringent regulations targeting the phaseout of small coal-fired boilers are now in effect in BTH and Shandong.

In the built-up areas of Beijing and Tianjin, and in the built-up urban area of the prefecture-level city and above in Hebei, existing coal-fired boilers with a steam production capacity of 35 tons per hour or below were forced to retire.

In the rural–urban fringe zones and the urban areas of other suburban districts, coal-fired boilers with a capacity of 10 tons per hour or below were normally phased out.

By the end of 2017, Tianjin, Hebei, and Shandong had gradually phased out on-site coal-fired boilers in production facilities for chemicals, papermaking, printing and dyeing, leathering, and pharmaceuticals.

Subsidies have been given to help accelerate the phaseouts.

Hebei incentivized the phase-out of coal-fired boilers with a subsidy of CNY30,000/ton of steam capacity, and likewise incentivized the replacement of those coal-fired boilers with new energy devices (e.g., heat pump) with a subsidy of CNY80,000/ton of steam.[9]

---

[9]   http://www.sohu.com/a/167466957_99931482.

Natural gas boilers and heat pumps were mandated as the main substitutes for coal-fired boilers. In some cases, the independent coal-fired boilers were allowed to be replaced by on-site natural gas heat and power plants combined.

These subsidy schemes have successfully facilitated most of the fuel switching from coal to national gas or electricity in the region.

Scaling Up Renewable Energy

Launched over 2 decades ago, a suite of policies in the PRC have been promoting renewable energy particularly in the power sector. As a result, the PRC now has the world's largest installed capacity of wind power and solar power, driven early on by a feed-in tariff. Making the national nonfossil fuel share target binding had elevated the role of renewables in provincial and municipal energy-system planning.

Under each Five-Year Plan, the share target had grown, while the feed-in tariff had been superseded by a feed-in premium. Renewable portfolio standards were announced in 2019.

Strengthening End-of-Pipe Control of Air Pollutant Emissions

End-of-pipe emissions control has been an ongoing measure in the PRC to decrease or eliminate the emission into the atmosphere of substances that can harm the environment or human health. Proving hugely popular for controlling air pollutants, the emissions reduction projects have been promoted on a large scale nationwide in the power sector, iron and steel sector, cement sector, and flat-glass sector.

End-of-Pipe Emissions Control in the Coal-Fired Power Sector

To control the power sector's air pollutant emissions, the PRC's thermal power plant emission standards have been revised multiple times to continually reduce the maximum allowable levels of flue gas[10] pollutants.

Under the Action Plan for Upgrading and Retrofitting of Coal-fired Power on Energy Conservation and Emission Reduction (2014–2020), new coal-fired power plants in the eastern region were required to reduce air pollutant emissions rates comparable to those released by gas turbines, while those in the central and western regions were encouraged to reduce air pollutant emissions rates comparable to those released by gas turbines.

By 2015, this standard was extended to all active-duty coal-fired units in the country. By 2020, coal-fired power plants that qualified for these retrofits were further required to achieve the PRC's ultra-low emissions standards, which had become stricter than those in developed countries such as the United States and the European Union (Table 3).

---

10   Flue gas is gas exiting to the atmosphere via a flue, which is a pipe or channel for conveying exhaust gases from a fireplace, oven, furnace, boiler, or steam generator. Quite often, flue gas refers to the combustion exhaust gas produced at power plants (Source: IEA. 2016. *World Energy Outlook Special Report: Energy and Air Pollution*. https://www.iea.org/reports/energy-and-air-pollution).

Table 3: Emission Permits for the Existing and New Power Plants
(mg/m³)

Emissions Standards	SO₂	NOₓ	PM
Ultra-low emissions standard	35	50	10
National standards (key areas)	50	100	20
National standards (general area/new build)	100	100	30
National standards (general area/existing)	200	100	30

Sources: (1) Emission standard of air pollutants for thermal power plants (GB 13223-2011); (2) National Development and Reform Commission, Ministry of Environmental Protection, and National Energy Administration. 2014. Action Plan for Upgrading and Transformation of the Energy Conservation and Emission Reduction of Coal-fired Power (2014–2020).

Modifications to comply with the ultra-low emissions modification in coal-fired power plants had been shown to remove 98.5%–99.4% of SO_2, 86.5%–93.1% of NO_X, and 98.9%–99.8% of $PM_{2.5}$. As of 2020, all coal-fired power plants in BTH and Shandong had completed instituting these upgrades.

End-of-Pipe Emissions Control for Boilers in the Industrial Sector

Coal-fired and natural gas-fired industrial boilers began facing tougher emissions standards for heat production. In 2015, the NEA required facilities with new coal-fired boilers that produce 20 steam tons per hour or more to install high-efficiency desulfurization[11] and dust removal facilities; in the case of new large boilers, the NEA required them to meet the same ultra-low emissions standards as those for electric power. By the end of 2018, coal-fired boilers in Tianjin, Hebei, and Shandong producing 65 steam tons or more per hour were required to complete these retrofits.

Natural gas boilers were required to achieve a NO_x emission concentration no greater than 50 mg/m³. Beijing had completed the low-nitrogen transformation of all gas-fired boilers by 2018. Tianjin, Hebei, and Shandong had basically completed retrofits on natural gas boilers by 2020.

Biomass boilers also started getting heavily regulated. Special low-emission boilers became mandatory, blending with coal and other high-emitting fuels was prohibited, high-efficiency dust removal facilities had to be installed, and implementation of ultra-low emissions standards was strongly encouraged.

---

[11]   Desulfurization means the removal of sulfur or sulfur compounds. In the field of air pollution control, it usually refers to the removal of sulfur dioxides. Sulfur dioxides are largely formed in the combustion of sulfur-containing fuels. The most used technology to remove SO_2 from the flue gas stream of combustion plants is the wet scrubber, which uses a calcium-based sorbent (typically limestone) to react with it and produce gypsum in a specifically designed vessel. Other technologies to control for SO_2 are spray dry scrubbers, sorbent injection, or a circulating fluid bed scrubber (source: https://www.merriam-webster.com/; IEA. 2016. *World Energy Outlook Special Report: Energy and Air Pollution.* https://www.iea.org/reports/energy-and-air-pollution).

End-of-Pipe Control of the Emissions in the Non-Power Sectors

As the major source of air pollutant emissions, the industrial sector had emission levels higher than those produced by electric power generation. As regulations tightened on the power sector, the industrial sector was therefore expected to emerge as the next frontier for end-of-pipe control.

In 2017, more than 20 industrial air pollutant emission standards were revised for the major high-polluting industrial sectors such as steel, cement, flat glass, coking, nonferrous metals, and brick manufacturing. This underscored an increasing focus on industrial pollution control by the Three-Year Action Plan to Win the Blue-Sky Defense War (2018).

Described in the following text are the key measures instituted for each industrial sector:

Iron and steel. The sintering process is a major emission-generating stage in iron and steel production. In the Ministry of Environmental Protection's latest ultra-low emission requirements, average hourly PM, SO_2, and NO_x emission concentration limits in the sintering process were 10 mg/m^3 of PM, 35 mg/m^3 of SO_2, and 50 mg/m^3 of NO_x, respectively.

As of 2015, the sintering machines with desulfurization equipment accounted for 80% of the total nationwide, and the Thirteenth Five-Year Plan required adoption of desulfurization with 98% efficiency at all facilities by 2020.

Future regulations would require the installation of denitrification[12] facilities and efficient dust removal techniques.

Cement. To meet the more stringent requirements for emissions reduction, cement production lines were required to implement additional end-of-pipe controls to achieve desulfurization, denitrification, and dust removal. The new requirements targeted reductions in SO_2 concentration in furnace flue gas to less than 100 mg/m^3, 50 mg/m^3, or even 20 mg/m^3.

Under the Thirteenth Five-Year Plan, compound desulfurization technology was required at facilities that rely on high-sulfur limestone.

Industrial boilers. Industrial boilers in the PRC are currently required to implement standard GB13271-2014, which is less stringent than the ultra-low emissions limits

---

[12]   Denitrification means the removal of nitrogen or nitrogen compounds. In the field of air pollution control, it usually refers to the removal of nitrogen oxides. Nitrogen oxides are any of several oxides of nitrogen, which are produced in combustion. Low nitrogen oxide burners and over-fire air are primary measures to abate NO_x emissions from combustion plants. Post-combustion nitrogen oxides control can be added to the primary measures to improve the effectiveness of nitrogen oxides control. Selective catalytic reduction and selective non-catalytic reduction are the most common technologies. The technologies treat the flue gas with a chemical reagent (mostly ammonia or urea) to react with the NO_x and form molecular nitrogen (source: https://www.merriam-webster.com/; IEA. 2016. *World Energy Outlook Special Report: Energy and Air Pollution.* https://www.iea.org/reports/energy-and-air-pollution).

for power plants. Implementation of the ultra-low emissions limits, which had already started in some areas, represents a major opportunity to further reduce emissions.

There had been some progress on this front. In 2018, these tougher limits were mandated in Hebei. In 2015, the emission limits issued by Beijing for SO_2 and particulate matter from coal-fired boilers (DB11/129-2015) were even lower than the ultra-low emission limits for coal-fired power plants.

Policies and Measures for the Residential and Services Sectors

The PRC's central and regional governments also moved to limit emissions from the residential and services sectors. However, the policies and measures to encourage the substitution of natural gas or electricity or both for coal in the regions raised strong concerns about affordability for low-income households and the possibility of natural gas supply shortages.

The 2018 Three-year Action Plan to Win the Blue Sky Defense War allowed for more flexibility to use other types of fuel. When necessary, heat could still be availed from electricity, natural gas, and even coal to ensure the provision of adequate safe heating services in the winter. For instance, residents in mountainous areas with limited alternatives were allowed to continue using coal in their homes, and the simultaneous use of less-polluting coal types was encouraged.

Ensuring natural gas supply was emphasized in the Three-year Action Plan to Win the Blue Sky Defense War. Priority was given to the residential and commercial sectors, especially in areas suffering from serious air pollution. At the same time, a moratorium was imposed through 2020 on new gas-fired power plants and natural-gas-fueled chemical projects, and the natural gas supply and pipeline capacity was expanded.

To incentivize substitution of natural gas and electricity for coal, a subsidy scheme was introduced in 2013. Suburban areas of Beijing were provided by the municipal government with a subsidy of as high as CNY12,000 for the installation of natural gas or electric heating. Rural facilities in Beijing that had to use electricity for space heating were granted tariffs at the lowest off-peak levels.

To limit overall demand, policies and measures to promote building energy efficiency also began to be implemented. These include energy performance standards for buildings and electrical appliances, subsidy schemes for energy savings, and a shift from heating charges from an area basis to a heat-volume basis.

Policies and Measures for the Transportation Sector

Fuel consumption standard for conventional vehicles. New light-duty vehicles were required to meet fuel consumption standards of 6.9 L/100 km by 2015 and 5 L/100 km by 2020, respectively, continuing a trend of increasing stringency. A stricter fuel consumption standard for heavy-duty vehicles was released shortly thereafter.

Vehicle pollutant emission standards. Tailpipe standards for air pollutants have grown increasingly strict over the years. Starting in 2013, Beijing implemented the Beijing 5/V emissions standards, which match the Euro 5/V emissions standards. In 2016, Tianjin, Hebei, and Shandong implemented the PRC's 5/V emissions standards, which were likewise equal to the Euro 5/V emissions standards.

The United Nations Environment Programme estimated that, in Beijing, the tightening of emission standards cumulatively contributed to the period between 1998 and 2017 up to 39% and 34% of the reductions in NO_x and primary $PM_{2.5}$ emissions compared with the uncontrolled scenario.

Phasing out of highly polluting vehicles. Several policies then targeted highly polluting vehicles. By 2015, so-called "yellow label" vehicles—gasoline vehicles with emission levels below the PRC 1 emission standards and diesel vehicles with emission levels below the PRC 3 emission standards—were phased out in Beijing and Tianjin; those in Hebei were phased out by 2017. The government has been subsidizing the scrappage of these vehicles.

Current driving restrictions now limit the use of high-pollution vehicles. For instance, since 2017, light-duty gasoline vehicles compliant only with the PRC 1 and the PRC 2 emission standards have been prohibited within the Fifth Ring Road in Beijing. Also, since 2017, heavy-duty diesel trucks that comply only with the PRC 3 emission standard or less have been prohibited within the Sixth Ring Road in Beijing.

Promoting the deployment of alternative fuel vehicles. For commercial, household, and industrial uses in the PRC, multiple policies now promote the production, purchase, and use of battery electric vehicles, plug-in hybrid electric vehicles, compressed natural gas vehicles, and liquefied natural gas vehicles.

BTH and Shandong are no exception; by 2020, they targeted a 70% share of alternative-fuel buses in bus fleets, while Tianjin targeted a 50% share. Also, by 2020, Hebei targeted an increase in its alternative-fuel vehicle population to more than 300,000 units. And in Shandong, some of the major cities like Jinan and Qingdao were required to replace their entire fleets with alternative fuel buses by 2020.

For freight vehicles, targeted for replacement with alternative fuel vehicles were categories that include vehicles for sanitation, postal, and logistics uses. In Beijing, e.g., postal vehicles, city express vehicles, and light sanitation vehicles below 4.5 tons by 2020 had to be replaced with electric vehicles. For private passenger vehicles, electric vehicle purchases are now subsidized, while restrictions on purchases of such vehicles—such as the requirement to enter auctions or lotteries—were made substantially lower.

For several years, both the central and local governments subsidized electric vehicle purchases; Table 4 shows the national subsidies for them over time. In 2013, however, the subsidy policy changed from battery-capacity-based to electric-range-based, such that the amount of subsidy started to be provided solely on the all-electric-range basis. The subsidy amount had also been declining each year.

Table 4: National Subsidies for Electric Vehicles
(CNY10,000)

Vehicle Type	All Electric Range	2013	2014	2015	2016	2017–2018	2019–2020
BEV[a]	$80 \leq R^c < 150$	3.5	3.15	2.8			
	$100 \leq R < 150$				2.5	2	1.5
	$150 \leq R < 250$	5	4.5	4	4.5	3.6	2.7
	$R \geq 250$	6	5.4	4.8	5.5	4.4	3.3
PHEV[b]	$R \geq 50$	3.5	3.15	2.8	3	2.4	1.8

Notes:
[a] BEV: battery electric vehicles
[b] PHEV: plug-in hybrid electric vehicles
[c] All electric range
Source: Various policies and guidelines on new energy vehicles from the Ministry of Finance of the PRC.

Adjustments to subsidies between 2017 and 2020 had taken into consideration both electric range and vehicle efficiency. Those subsidies are now currently being phased out because of the diminishing cost of replacement technology and growing constraints on government budgets.

Optimization of transportation modes. Expansion of public transportation and of the rail network in BTH and Shandong has contributed to increased mobility while limiting environmental impact.

Urban rail transit in Beijing reached 727 km in 2020. Tianjin's system alone targeted to reach 375 km. Also, by 2020, Hebei's railway mileage within the province was targeted to exceed 8,500 km, including 2,000 km of high-speed railway and 80 km of urban rail transit.

Vehicle purchase and use restrictions. License plate lotteries, launched in Beijing in 2011 and in Tianjin in 2014, sought to limit the rapid growth of vehicle stock that had been contributing to congestion in the region.

Quotas were adjusted dynamically. In Beijing, the quota was tightened from 240,000 vehicles/year in 2011 to 150,000 vehicles/year in 2014, and to 100,000 vehicles/year in 2018. Electric vehicles were not subject to quotas.

To decrease vehicular congestion and air pollution, some regions have introduced restrictions based on the final digit of a vehicle's license plate number; in Beijing and Tianjin, a vehicle cannot be driven 1 day per week.

Carbon Emissions Trading Pilot Programs in Beijing and Tianjin

Carbon emissions trading has been under development in the PRC for almost a decade. In 2011, the NDRC selected Beijing, Tianjin, Shanghai, Shenzhen, Hubei, and Guangdong for CO_2 emissions trading pilots that were launched in 2013–2014. The pilots covered firms that contribute 40% of total CO_2 emissions in Beijing and 60% in Tianjin.

Pilot systems had set the thresholds at the firm level. In Beijing, included in the emissions trading system were firms emitting over 10,000 tons of CO_2 in 2013 and 2014, and from 2015 onward, those with emissions of over 5,000 tons of CO_2 were likewise included (Table 5). The revised threshold caused the number of firms covered to rise from 415 in 2013 to 943 by 2017. The number slightly decreased to 859 in 2020 as some firms suspended production due to industrial restructuring or other reasons.

In Tianjin, included in the emissions trading system were firms emitting over 20,000 tons of CO_2. The number of covered firms remained relatively stable, with the 114 firms covered in 2014 staying close at 109 in 2016.

As indicated in Table 5, the Beijing system has a more comprehensive sectoral coverage.

Table 5: Thresholds and Coverage of the Carbon Emissions Trading Pilot Programs in Beijing and Tianjin

Pilots	Thresholds	Number of Covered Firms	Covered Sectors (2020)
Beijing	Annual Emissions ≥ 10,000 ton CO_2 (2013–2014) Annual Emissions ≥ 5,000 ton CO_2 (2015–2020)	415 (2013) 543 (2014) 551 (2015) 947 (2016) 943 (2017) 903 (2018) 843 (2019) 859 (2020)	Power, heat, cement, petrochemical, other manufacturing, services, transportation (including civil aviation)
Tianjin	Annual Emissions ≥ 20,000 ton CO_2	114 (2014) 109 (2016) 109 (2017) 107 (2018) 113 (2019) 104 (2020)	Power, heat, building materials, papermaking, iron and steel, chemical, petrochemical, oil and gas exploitation, aviation

Source: Tsinghua University.

The total trading volumes are shown in Table 6. From December 2013 to June 2021, CO_2 prices in Beijing ranged from about CNY30/ton to about CNY100/ton. In Tianjin, the range over the same period fluctuated between CNY7/ton and CNY48/ton.

The PRC's national carbon emissions trading system entered into operation on 16 July 2021. The allowance price is CNY48/ton for the first transaction. The average transaction price was about CNY45/ton by October 2021, with the highest transaction price being CNY61.07/ton, and the lowest transaction price being CNY41/ton.

Table 6: Allowance Trading Volume
(10,000 ton CO_2)

Year	Beijing	Tianjin
2014	106	99
2015	126	53
2016	242	31
2017	238	116
2018	306	0.07
2019	311	4.38
2020	114	520

Source: Carbon Emissions Trading. https://www.sohu.com/a/458489924_473133 (accessed 3 May 2021).

Limitations to Current Measures and Policies

Our review of the PRC's current policies and measures revealed a wide variety of policy instruments and growing ambition of target goals over time. However, a main shortcoming seen in the current approach was arguably the lack of an integrated plan to achieve broader air quality and climate goals.

For air quality, the current policies have not clearly charted a path to achieve the National Ambient Air Quality Standards, nor have they considered interactions with climate mitigation efforts that focus on reducing CO_2 from many of similar intended activities. The climate mitigation efforts also bore no clear relationship to the global 2°C target. Moreover, these measures were strongly focused on the near term, whereas targets with a longer-term horizon could have allowed affected industries to plan way ahead. Especially, targets should be established through 2035 and be aligned with the PRC's pathway to achieve CO_2 emissions peak before 2030 and to build a Beautiful China[13] by 2035. The remainder of the study thus puts a strong focus on identifying policies that can more capably achieve air quality and climate change goals in an integrated and much more cost-effective manner.

---

[13]   According to the long-term strategies unveiled at the 19th National Congress of the Communist Party of the PRC, the period from 2020 to 2035 is the first stage to build a modern socialist PRC. A major improvement in the environment is planned, with the basic goal of having built a Beautiful China by 2035.

4 Methodology and Alternative Scenarios

Modeling Approach

To simulate alternative scenarios for BTH and Shandong through 2035, the Regional Emissions Air-Quality Climate and Health (REACH) framework was used. The REACH framework combines an energy-economic model, an emissions inventory model, an air quality model, and a health effects evaluation model. REACH simulates the complex and nonlinear actions among economic activity, energy use, CO_2 emissions, air quality, and health, capturing projected changes at the provincial level and over time under alternative policy scenarios.

In this new study, the REACH framework was used in the air quality and climate policy analysis under different underlying modeling suites (Li et al., 2018).[14] Core elements of the constituent models were updated and improved for the study.

Figure 6 shows the relationship among the components of the REACH framework, which uses the energy-economic model known as the China Regional Energy Model (C-REM). This model can rationalize the PRC's economy and energy system at the provincial level by simulating the impact of projected production, consumption, the inter-provincial and international level of energy use, and CO_2 emissions.

In the REACH framework used in this new study, for instance, policies can change the relative prices of goods and services as a function of their air pollution or CO_2 intensity. This can help bring about desired levels of production and consumption that would minimize energy use and the resulting pollutants. Projections for CO_2 and air pollutant emissions by province are obtained by multiplying energy use with the fuel-specific emissions factors in each policy scenario.

Based on a detailed emissions inventory for the PRC, these projections are assigned spatially by region. The air quality model then uses the spatially distributed air pollutant emissions data as an input to simulate concentrations across space and time.

To calculate the monetized health impact of air pollution in each region, the C-REM has an embedded health effects evaluation module that uses air pollution concentration data as an input. These monetized impacts are then fed back into the energy-economic (C-REM) model to enable the REACH framework to reflect the indirect costs or benefits of variations in health damage caused by air pollution.

---

[14] M. Li, et al. 2018. Air Quality Co-benefits of Carbon Pricing in China. *Nature Climate Change*, 8(5), p. 398.

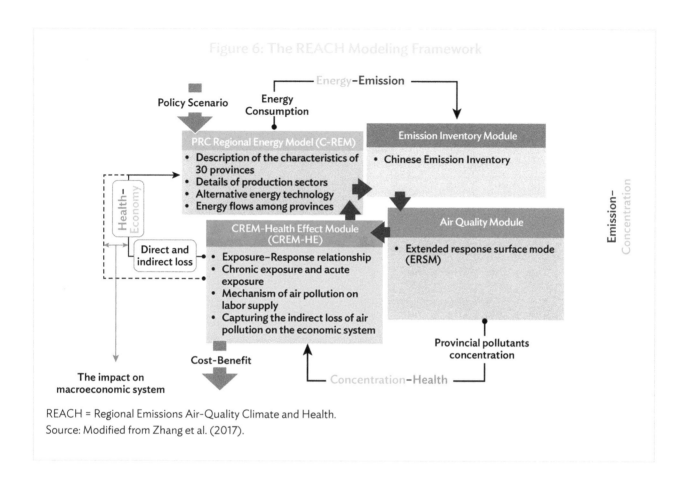

Figure 6: The REACH Modeling Framework

REACH = Regional Emissions Air-Quality Climate and Health.
Source: Modified from Zhang et al. (2017).

Appendix 2 provides more details on the structure of the REACH framework.

Compared to prior applications of the REACH framework, the version used for this analysis features the following six improvements:

(i)   updating of the C-REM's benchmark provincial input–output data with the latest available (2012) to replace the 2007 benchmark data used in earlier versions;

(ii)   disaggregation of 13 production sectors in the original version into 21 production sectors within each region to better rationalize energy-intensive industries;

(iii)   adoption of a newer version of the PRC's emissions inventory that has a more detailed resolution of regions, pollutants, and technologies;

(iv)   calibration of diffusion possibilities for end-of-pipe control technologies by sector to reflect prior adoption status;

(v)   updating of the module to capture the health impacts of air pollution; and

(vi)   taking into account the uncertainties associated with different health valuation methods.

Features of the REACH Model

The advantages of the REACH modeling framework are described in more detail in the following text. The goal is to justify the selection of this modeling framework, discuss its limitations and caveats, and describe how components were adjusted to represent BTH and Shandong.

Propagation of Policy Impacts across Sectors

By incorporating the C–REM energy-economic module, the REACH framework was enabled to capture the impacts of policies on supply, demand, and relative prices across sectors and regions. This is because, as a general equilibrium model, C–REM has capabilities superior to the vast majority of energy-economic models that simulate the effects of environmental policies.

Other widely used models assume end-use demand that does not respond to changes in relative prices and limit policy impacts only to targeted sectors. In contrast, the C–REM provides greater realism in capturing these feedbacks and the effects of policy on prices of key production inputs such as labor, capital, and energy. These sectoral interactions are especially important when considering economy-wide climate policies of the type modeled in this analysis.

In C–REM, the policies that can be represented range from command-and-control policies that directly constrain the quantity or efficiency of energy use, to market-based policy instruments such as carbon pricing on fuels, or to the promotion of the specific energy technologies. In C–REM, policies act through changes in the relative prices of goods as economic activities move to a new equilibrium that meets policy constraints at least cost.

For example, a policy that puts a price tag on CO_2 emissions would induce substitution away from CO_2-intensive activities. When the cost of reducing CO_2 in a sector is high, sector activity would decrease in response to the additional cost burden. In contrast, sectors that face a wider range of substitution possibilities with limited incremental costs (e.g., renewable energy in electric power) would shift to cleaner modes of production that would limit the overall adverse impact.

Changes in CO_2 emissions under each policy scenario would follow directly from changes in energy use, and the CO_2 emissions themselves would then be calculated using fuel-specific emissions factors.[15] However, one limitation is that while the study reports the provincial CO_2 emissions from domestic fossil-fuel combustion, it does not consider those from net imported electricity from outside the region.

---

[15] CO_2 emission factors of coal, oil, and natural gas in the study are 2.66 ton/tce, 1.73 ton/tce, and 1.56 ton/tce, respectively.

Modeling Air Pollutant Emissions

After each run of the C-REM, corresponding changes in the emissions of air pollutants are computed using technology-specific, energy-based emissions factors. They are further adjusted to account for end-of-pipe emissions controls.

The model implicitly captures the costs of reducing air pollutant emissions through output reduction, sectoral composition effects, and fuel switching. However, assumptions about the application of end-of-pipe technology need to be applied separately to the C-REM outputs, for the C-REM model is at present unable to reflect the costs of adopting end-of-pipe control technologies.

Projecting Air Quality

Air quality outcomes are a product of complex, nonlinear interactions among precursor[16] pollutants, which could be projected over time using an atmospheric chemistry[17] model. Inputs are spatially distributed levels of pollutants that are specified in the scenario-specific emissions inventory.

The response surface model (RSM) is used to simulate air quality outcomes. For a representative month in each scenario, the model would simulate the interactions among pollutant emissions, the weather variables such as temperature, precipitation, and humidity, and the background atmospheric components.

Modeling Health Effects

By making a projection of spatially distributed air quality, it becomes possible for REACH to simulate health effects on populations. The approach used in the study therefore involved first computing the exposure to air pollution under each scenario. Next, an exposure–response function for determining the associated mortality was used to capture differences across provinces and age groups.

Each death was valued using a value of a statistical life (VSL) specific to the PRC. This was in addition to valuing the so-called direct impact on mortality of each policy scenario and the indirect effect these losses would have on the PRC's economy.

The version of C-REM used in the study had been augmented with a substitution factor for health-services needs and leisure consumption; that factor puts a specific

---

[16]   Precursor is a compound that participates in a chemical reaction that produces another compound (Source: UNEP. 2019. *Air Pollution in Asia and the Pacific: Science-based Solutions.* https://www.ccacoalition.org/en/resources/air-pollution-asia-and-pacific-science-based-solutions-summary-full-report).

[17]   Atmospheric chemistry is the study of the chemical composition and transformations of the natural planetary atmosphere. It focuses upon understanding natural and anthropogenic emissions to the atmosphere, the transport, chemical, and physical transformations of atmospheric constituents, and the effects of air pollution and atmospheric chemistry on the environment and, particularly, on human health (Source: UNEP. 2019. *Air Pollution in Asia and the Pacific: Science-based Solutions.* https://www.ccacoalition.org/en/resources/air-pollution-asia-and-pacific-science-based-solutions-summary-full-report).

value to both lost labor and lost leisure time caused by the degradation of air quality. The health impacts are then reflected in an extension of the social accounting matrix, which tracks by province the monetized impacts of these losses over time.

Reflecting Regional Differences

Within the C-REM, special attention was given to the representation of Beijing, Tianjin, Hebei, and Shandong, considering that variables about these provinces had already been captured by the overall data that were gathered to represent the PRC's 30 provinces in the C-REM.

The C-REM model also captured the impact of trade among provinces, including their usage of renewable energy, as well as the effects of air pollutant movement across provincial borders.

Designing the Scenario Exercise

As described in the previous chapter, the PRC government has applied numerous policy levers of varying stringency to control air pollution and CO_2 emissions. In this respect, the objective was to find and evaluate the most appropriate combinations of these levers that can deliver the best policy packages for meeting the country's air quality and climate change mitigation goals.

As described in Table 7, a total of six scenarios were projected that apply two broad categories of interventions across BTH and Shandong, namely energy efficiency and fuel switching, and end-of-pipe control technologies.

Most of the energy efficiency and fuel-switching measures are meant to contribute to the elimination or reduction of both CO_2 and air pollutants. For instance, the power sector would present a great opportunity for co-control when nonfossil generation replaces the usage of coal.

In contrast, most end-of-pipe control measures address only the problem of air pollutants without considering possible synergies into account. In the case of transport, e.g., measures to control tailpipe NO_x have no effect on CO_2 emission reductions. This makes these control measures expensive owing to the limited potential to substitute away from fossil fuels and the expected continued growth in the transportation demand.

On the other hand, switching to electric vehicles would limit the rise in CO_2 emissions to the extent that the generation of electricity uses a clean, non-CO_2-generating process.

To project the impact of each scenario's policies on the economy of the PRC, a set of assumptions and constraints in the C-REM were imposed. Whenever possible, policies used in the C-REM were modeled in a manner that captures both the impact on the costs and health benefits.

Table 7: Simulated C-REM Scenarios

Scenario	Energy Efficiency and Fuel Switching			End-of-Pipe Control	Reported as
	Limit of CO_2 Emission	Renewable Energy	Electric Vehicle		
Scenario 1	NDC/4.5% of annual average CO_2 intensity reduction	Continued efforts	Continued efforts	Continued efforts	Current Policies scenario
Scenario 2	NDC/4.5% of annual average CO_2 intensity reduction	Continued efforts	Continued efforts	Best available technologies	
Scenario 3	Enhanced/5.5% of annual average CO_2 intensity reduction	Enhanced efforts	Enhanced efforts	Continued efforts	
Scenario 4	Enhanced/5.5% of annual average CO_2 intensity reduction	Enhanced efforts	Enhanced efforts	Best available technologies	
Scenario 5	2°C /6.5% of annual average CO_2 intensity reduction	Enhanced efforts toward 2°C	Enhanced efforts	Continued efforts	
Scenario 6	2°C /6.5% of annual average CO_2 intensity reduction	Enhanced efforts toward 2°C	Enhanced efforts	Best available technologies	Environmental goals attainment scenario

C-REM = China Regional Energy Model, NDC = Nationally Determined Contribution.
Source: ADB (compiled by authors).

Energy Efficiency and Fuel Switching

The energy-efficiency and fuel-switching measures were simulated primarily through economy-wide targets for CO_2 intensity reduction. Three emission trajectories were simulated: (i) CO_2 emissions peaking in 2030 consistent with the PRC's NDC; (ii) enhanced efforts to reach peak emissions in 2025; and (iii) stronger and even more intensive efforts to meet the 2°C temperature control limit on postindustrial temperature rise set by the 2015 Paris Climate Accord.

Imposed on these trajectories are several policy interventions, among them an emissions trading system to achieve an absolute limit on CO_2 emissions, a renewable energy promotion, and an electric vehicle promotion.

Upper Limits on CO_2 Emissions

The upper limits on CO_2 emissions in the region were primarily simulated using economy-wide targets for CO_2 intensity reduction.

In Scenarios 1 and 2 for BTH and Shandong,[18] CO_2 intensity reductions of 4.5% per year between 2015 and 2035 were simulated in the region to achieve a peak in 2030 consistent with the NDC at the national level.

In Scenarios 3 and 4, CO_2 intensity reductions of 5.5% per year between 2015 and 2035 in the region were simulated to achieve the CO_2 emission peak in 2025 at the national level.

In Scenarios 5 and 6, CO_2 intensity reductions of around 6.5% per year between 2015 and 2035 in the region were assumed. This nationwide CO_2 emission trajectory reflects the PRC's committed contribution to limiting the global temperature increase to well below 2°C during this century in line with the projections of national CO_2 emission trajectories of related studies.

It is clear that mitigating climate change would require severely limiting and eventually reducing absolute levels of CO_2 emissions. Based on the targeted CO_2 price, the study, therefore, set the following regional upper limits of CO_2 emission: 1.74 billion tons by 2025, 1.53 billion tons by 2030, and 1.27 billion tons by 2035.

Support for Renewable Energy

To represent different levels of policy effort, the renewable energy deployments by province in the C-REM were modeled by adjusting subsidies to renewable energy and the resource limits of renewable energy. To integrate more bottom–up information, the C-REM was coupled with the following models:

(i)   a bottom–up model—in this case the Renewable Electricity Planning and Operation (REPO) model—that simulates the deployment of power generation technologies in 30 provinces in the PRC; and

(ii)   the China-in-Global Energy (C-GEM) model that focuses on the energy economic analysis at the national level.

Figure 7 shows the interactions of these three models.

First, the projections of national GDP and CO_2 emissions from 2015 to 2035 were inputted from C-GEM to C-REM to calibrate the provincial GDPs and the upper limits of provincial CO_2 emissions. Second, the simulated provincial electricity demand in C-REM and the CO_2 emissions of the power sector in C-GEM were inputted to the REPO as constraints to generate the provincial power generation by technology. Third, the C-REM then used the provincial power generation data from REPO to calibrate its provincial power generation data.

---

[18]   It is assumed that the national emission reduction target would continue to be disaggregated to the provincial level, and the provincial CO_2 intensity targets would be divided into five categories through 2035 in line with the national Thirteenth Five-Year Work Plan on Greenhouse Gas Control. BTH and Shandong would still adopt the most stringent target of carbon intensity reduction—along with Shanghai, Jiangsu, Zhejiang, and Guangdong.

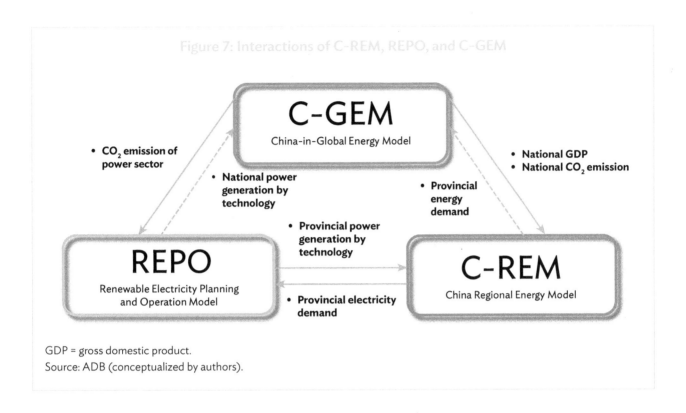

Figure 7: Interactions of C-REM, REPO, and C-GEM

GDP = gross domestic product.
Source: ADB (conceptualized by authors).

Electric Vehicles

Each scenario simulated different levels of policy effort to promote electric vehicles. Electric vehicle deployment targets were achieved by adjusting subsidy levels until the share of spending on them became consistent with fleet targets.

To simulate the costs associated with scaling the production of electric vehicles, an input factor was parameterized in the same manner as the input factor used to represent adjustment costs associated with economic restructuring. This factor represents the additional costs, such as range anxiety, that are associated with moving to production at scale and with having to overcome adoption barriers.

Policy targets simulated in the Current Policies scenario were designed to make the penetration rate of electric passenger vehicles in 2035 reach 17.2% in Tianjin, Hebei, and Shandong, and reach 28% in Beijing by that same year.

In the EGA scenario, the policy targets simulated the penetration rate of electric passenger vehicles in Tianjin, Hebei, and Shandong by 2035 to reach 32.2%, and in Beijing to reach 60%.

End-of-Pipe Control Measures

The scenarios adopted two different assumptions about the penetration of end-of-pipe control measures. First, the continued efforts measure package (EOP-CE) assumed that all existing policies would continue to be implemented but with no new control measures imposed. Second, it was assumed that the best-available-technologies measure package (EOP-BAT) would be used to identify the state-of-the-art emissions control technologies that would be widely adopted in the future.[19]

Pollution control requirements are modeled by adjusting emissions factors directly, so they do not reflect the costs of installing pollution-control equipment. The end-of-pipe control measures considered in the study thus provided for the following: (i) ultra-low emissions technologies in the power sector; (ii) pollution control and upgrading in key industries; (iii) ultra-low emission transformation for industrial coal-fired boilers; (iv) limiting emissions from the building sector; and (v) fuel quality improvement and emission standards upgrading for vehicles.

Ultra-Low Emissions Technologies in the Power Sector

In all scenarios, large coal-fired power plants (300 MW of noncirculating bed boilers) should aim to complete ultra-low emission modifications by 2020. As a result, the average desulfurization and denitrification efficiency of all coal-fired units that have completed ultra-low emission modifications should by then have already reached 98.5% and 90.0%, respectively.

Under EOP-CE, it was assumed that the desulfurization and denitrification efficiency would increase by 2035 to 98.8% and 91.0%, respectively, and that under the EOP-BAT package, this efficiency would further increase to 99.0% and 91.5% by 2035, respectively.

For dust removal, it was assumed that by 2020, there would be no more need for traditional removal of untreated electric dust, and that by 2035, high-efficiency dust removers like the electrostatic-fabric integrated precipitator (ESP-FF) and fabric filter (FF) would already be widely applied for the purpose. See Appendix 1 for the details of these end-of-pipe control assumptions.

Pollution Control and Upgrading in Key Industries

In each of the scenarios, the pollution control assumptions reflected current and possible future efforts to apply advanced end-of-pipe control technology. The focus was on industries targeted for upgrading under the ultra-low emissions standards.

In iron and steel, all scenarios assumed that by 2030, 100% of the denitrification facilities would have already been installed and operational. The difference between the EOP-CE and EOP-BAT packages was that the latter assumed a broader application of highly efficient selective catalytic reduction (SCR) denitrification technology.

In cement, scenarios differed in the extent of desulfurization facilities. For denitrification, it was assumed that the most widely used selective non-catalytic reduction (SNCR) denitrification technology would be applied. It was also assumed that some production lines would use low nitrogen burners, which would further

---

[19]   X. Qin. 2018. *Report on the Work of Government of Hebei.* (2020-10-06) [2018-02-05]. http://www.hebei.gov.cn/hebei/14462058/14471802/14471805/14867265/index.html.

improve denitrification efficiency. To further reduce dust emissions, ESP-FF was simulated to replace FF, which is already being used in most cement production lines.

In flat glass, the EOP-CE and EOP-BAT packages differed in that the latter assumes a broader application of High-Efficiency Flue Gas Desulfurization (HEFGD) for desulfurization, of SCR for denitrification, and of bag filters and electric-bag filters with wet electrostatic precipitators (WESPs) for dust removal. Appendix 1 details the technology assumptions for each of these scenarios.

Other than the above key industrial sectors, the refined aluminum, coking, brick-making, lime, petroleum refining, sulfuric acid, and nitric-acid sectors have likewise considered the adequacy of their end-of-pipe control technologies and are now likewise upgrading them.

Ultra-Low Emission Transformation for Industrial Coal-Fired Boilers

As regulators strictly implement the ultra-low emissions standard for industrial coal-fired boilers, the use in industry of highly efficient end-of-pipe control technologies was assumed to greatly increase across all scenarios. Appendix 1 provides detailed assumptions about the timing of the rollout.

Limiting Emissions from the Building Sector

In the building sector, the diverse and dispersed pollutant emissions could not be adequately reduced with end-of-pipe control equipment alone. The scenarios therefore assumed the availability of other strategies for abating emissions, such as the use of briquette and clean stoves.

As mentioned in the previous section, some coal use for heating in remote mountainous areas would be allowed in subsequent years, and measures to promote briquette and clean stove use would be implemented in these areas.

In the EOP-BAT package, all bulk coal demand for heating or cooking by 2030 would be completely substituted by briquette or advanced clean stove use; in the EOP-CE package, the substitution ratio by 2030 would only be 50%.

Appendix 1 provides more details for these packages.

Fuel Quality Improvement and Emission Standards Upgrade of Vehicles

In the coming decades, the tightening of fuel quality and tailpipe emissions standards would be implemented nationwide in the PRC to help achieve the lowest possible levels of emissions worldwide.

The implementation of very strict rules for the continuing reduction in motor vehicle tailpipe emissions is expected following the current specified schedules. All the scenarios thus assumed the enforcement of fuel quality standards and tailpipe emissions standards for all light vehicles, heavy diesel vehicles, heavy gasoline vehicles, motorcycles, internal combustion locomotives, and inland vessels.

Two Focal Scenarios

Among the six scenarios, Scenarios 1 and 6 have been chosen as the "representative" scenarios for BTH and Shandong for evaluating the costs and health benefits targeted by the increasingly ambitious but feasible policy packages for air-quality improvement and climate change mitigation.

The two representative scenarios are defined as follows:

Current policies scenario (Scenario 1): This scenario projects in the simulation the previous policies still in effect. In this scenario, the pace of CO_2-intensity reduction targeted by the government in the Thirteenth Five-Year Plan (2015–2020) would be maintained through 2035 in keeping with the PRC's Nationally Determined Contribution under the Paris Agreement. End-of-pipe control of air pollutant emissions would be strengthened gradually under this scenario.

Environmental goals attainment scenario (Scenario 6): This scenario projects in the simulation the development of the economy under policies that limit $PM_{2.5}$ concentrations in the four municipalities and provinces to no more than $35\,\mu g/m^3$ by 2035, and reduce CO_2 emissions in line with the 2°C temperature control target.

The emission reductions would be accomplished through these two measures: (i) by requiring energy efficiency and fuel switching, and (ii) by mandating installation and operation of stringent end-of-pipe air pollution controls. Appendix 1 provides the detailed information about each of these measures.

Attaining the ambient air quality targets proved difficult in modeling the initial set of scenarios for the study. Thus, the scenario design had to be conducted iteratively or repeatedly, using stronger assumptions each time until the scenario achieved the target using a balanced combination of policy measures. The only scenario that was able to fully attain the environmental goals is the EGA scenario, which will be described next in more detail in this report.

5

Scenario Results

What follows are the simulated outcomes of the Current Policies scenario and the EGA scenario. The outcomes presented are the resultant effects or changes in economic activity, in energy use, in emission levels, in air quality, and in human health. A summary of the costs and benefits of implementing the EGA scenario is then presented in comparison to those of implementing the Current Policies scenario.

Energy System Transformation

The package of policies introduced by the Current Policies and EGA scenarios remarkably improved the regional energy system in several ways. They particularly demonstrated that the reduction of coal usage is crucial to achieving the PRC's climate change and air quality goals. Both the Current Policies and EGA scenarios demonstrably exhibited the need for large reductions in coal use. Figure 8 shows the energy consumption per province.

In 2015, coal use was 8 Mtce in Beijing, 41 Mtce in Tianjin, 236 Mtce in Hebei, and 290 Mtce in Shandong, respectively. The share of coal use in the total energy consumption of BTH and Shandong was about 70%, which is higher than the national level (63.8%) at that time.

In the EGA scenario, coal use by 2025 would fall to no more than 0.7 Mtce in Beijing, to 30 Mtce in Tianjin, to 215 Mtce in Hebei, and to 272 Mtce in Shandong. By 2035, coal use would decrease to no more than 0.5 Mtce in Beijing and would fall to 14 Mtce in Tianjin, to 146 Mtce in Hebei, and to 181 Mtce in Shandong, as shown in Figure 8. The share of coal use in the total energy consumption of BTH and Shandong would fall to 38% by 2035, which is almost equal to the national level at that time in the 2°C scenario projected by related studies.[20]

Both the Current Policies and the EGA scenarios also exhibited the need for a substantial increase in renewable energy utilization from the preexisting levels.

In 2015, the proportion of renewable energy in the electricity supply of each province was 4% in Beijing, 2% in Tianjin, 9% in Hebei, and 5% in Shandong. These proportions were far below the national average of 24% for that year.

---

[20] T. Wang. 2019. *Emission Co-Mitigation Pathways of CO$_2$ and Air Pollutants of China.* Tsinghua University.

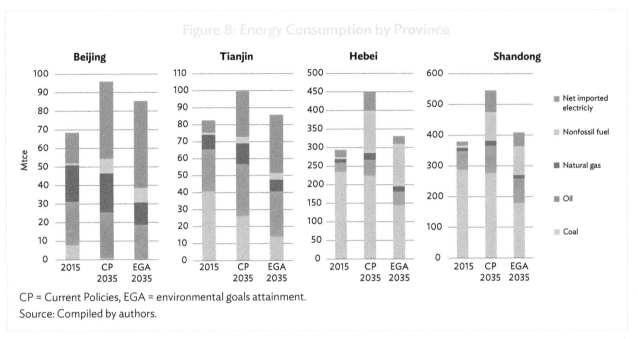

Figure 8: Energy Consumption by Province

CP = Current Policies, EGA = environmental goals attainment.
Source: Compiled by authors.

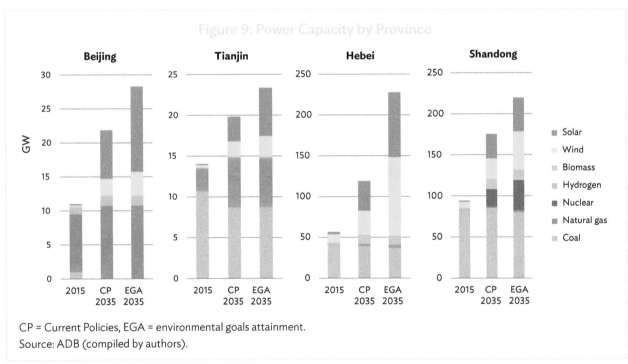

Figure 9: Power Capacity by Province

CP = Current Policies, EGA = environmental goals attainment.
Source: ADB (compiled by authors).

By 2017, all of Beijing's coal-fired power plants had already closed. In the EGA scenario, installed capacity of wind power from 2015 to 2035 would grow from 200 MW to 3.5 GW in Beijing, from 320 MW to 2.5 GW in Tianjin, from 10 GW to 96.5 GW in Hebei, and from 7.2GW to 47 GW in Shandong.

Over the same period, use of solar power was simulated to increase in 2035 from 165 MW to 12.5 GW in Beijing, 125 MW to 5.9 GW in Tianjin, 2.8 GW to 79 GW in Hebei, and from 1.3 GW to 41 GW in Shandong. The projections considered the land use requirement for solar photovoltaics in the four subregions, especially the two municipalities, i.e., Beijing and Tianjin. Suitable roof areas have been estimated and taken into account (Figure 9).

Along with the remarkable increase in the utilization of local renewable energy, the electricity imports to the four provinces would significantly increase. In 2035, the total electricity imports would grow to 480 TWh, accounting by then for about 27% of the region's electricity demand.

CO$_2$ and Air Pollution Emissions

Emission levels significantly dropped under the simulated policies and measures that worked through two channels of improvement, namely energy efficiency and fuel switching, and end-of-pipe control.

Under both the Current Policies scenario and the EGA scenario, the CO$_2$ emissions in the region had plateaued at around 1.83 billion tons starting in 2015. In the Current Policies scenario, CO$_2$ emissions by 2025 would peak at 1.94 billion tons and decrease slowly by 2035 to 1.84 billion tons.

In the EGA scenario, the CO$_2$ emissions by 2025 would further decrease by 6% to 1.72 billion tons and further go down by 2035 by 32% to 1.25 billion tons. The region would achieve CO$_2$ emissions peak before 2025, earlier than what have been planned for the nationwide (Figure 10).

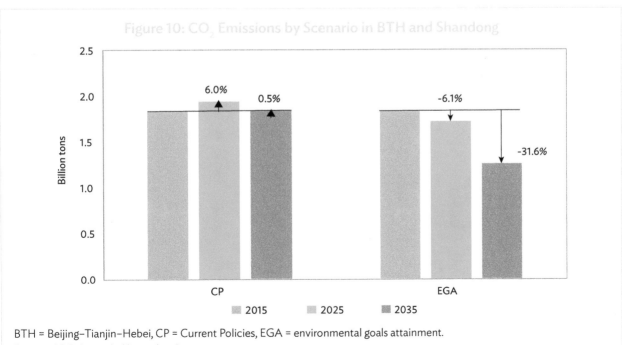

Figure 10: CO$_2$ Emissions by Scenario in BTH and Shandong

BTH = Beijing–Tianjin–Hebei, CP = Current Policies, EGA = environmental goals attainment.
Source: ADB (compiled by authors).

Under the EGA scenario, the simulated emissions of SO_2, NO_x, and primary $PM_{2.5}$ would decrease sharply from their respective 2015 levels. Shown in Figure 11 are the reductions that would be achieved by the EGA scenario for each emissions type by 2035 relative to 2015.

Under the EGA scenario, the SO_2, NO_x, primary $PM_{2.5}$, and CO_2 emissions in the industrial sector by 2035 relative to 2015 would decline by about 85%, 63%, 75%, and 39%, respectively; in the power sector by 85%, 80%, 83%, and 34%, respectively; in the transport sector by 14%, 81%, 79%, and –18%, respectively; and in the services and household (S&H) sector by 63%, 59%, 81%, and 16%, respectively.

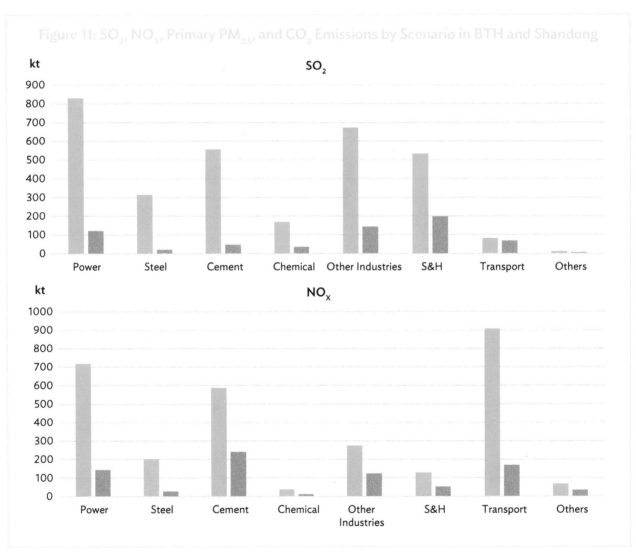

Figure 11: SO_2, NO_x, Primary $PM_{2.5}$, and CO_2 Emissions by Scenario in BTH and Shandong

continued on next page

Figure 11 *continued*

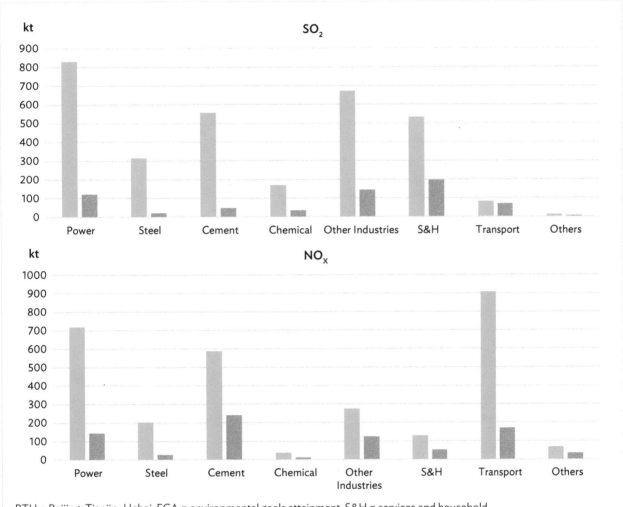

BTH = Beijing–Tianjin–Hebei, EGA = environmental goals attainment, S&H = services and household.
Notes: (i) Steel sector includes iron and steel and nonferrous industries. (ii) Cement sector refers to the nonmetallic industries.
(iii) Chemical sector includes chemistry and refined oil industries. (iv) Other industry sectors include mining, construction,
transport equipment, food, textile, electronic equipment, machinery manufacture, water, etc.
Source: ADB (compiled by authors).

The total emissions reduction of SO_2, NO_x, primary $PM_{2.5}$, and CO_2 by 2035 relative
to 2015 would be 2.52 Mt, 2.12 Mt, 0.96 Mt, and 576 Mt, respectively. The industrial
sector would contribute most of the reductions in SO_2 emissions (58%), primary
$PM_{2.5}$ emissions (55%), and CO_2 emissions (54%); the transport sector would
contribute most of the reductions in NO_x emissions (35%).

Under the EGA scenario for 2035 relative to 2015, significant SO_2, NO_x, primary
$PM_{2.5}$, and CO_2 emissions reductions at the provincial level would occur in all four
provinces. Especially, Hebei and Shandong would contribute substantial declines in
absolute terms (Figure 12). In both of these provinces, with the growing stringency of
the climate and air quality policies, the industrial sector and the power sector would

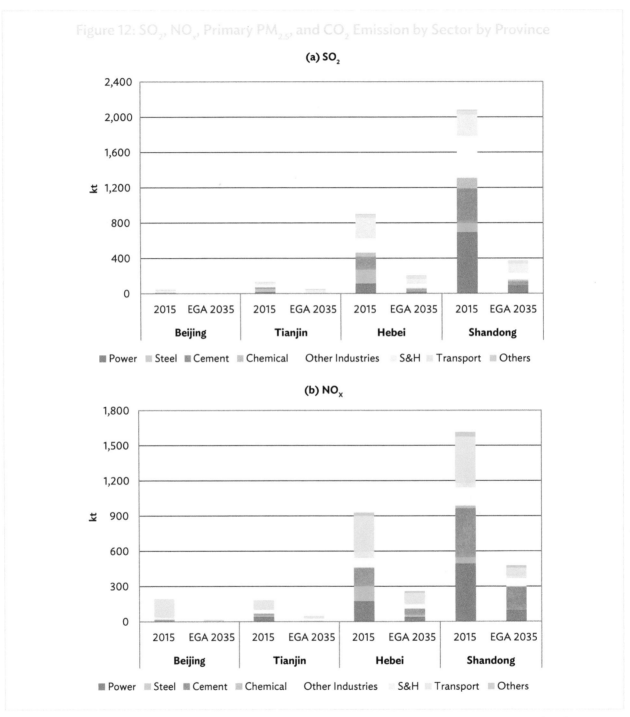

Figure 12: SO$_2$, NO$_x$, Primary PM$_{2.5}$, and CO$_2$ Emission by Sector by Province

continued on next page

Figure 12 *continued*

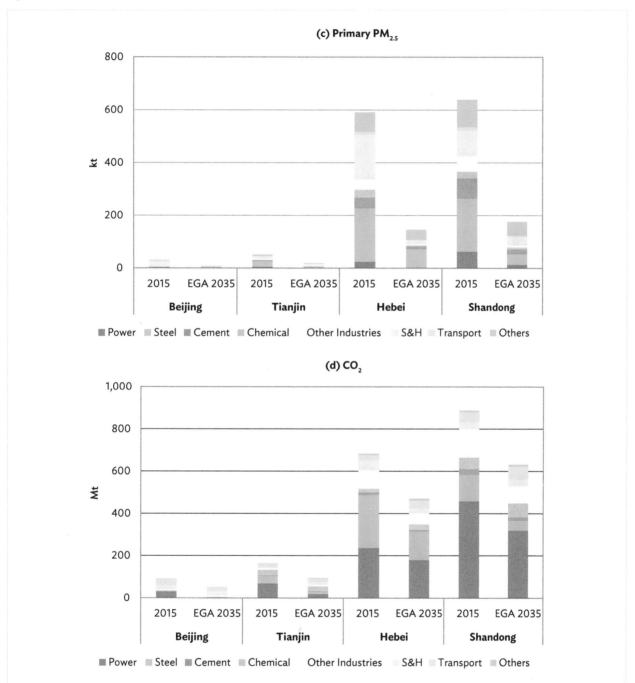

EGA = environmental goals attainment, S&H = services and household.
Source: ADB (compiled by authors).

contribute the most of all four kinds of emission reductions, reflecting abundant low-cost opportunities to reduce emissions in the provinces with a large share of energy-intensive sectors. For Beijing, the SO_2 and primary $PM_{2.5}$ emissions are already relatively low; nonetheless, their decline potential in relative terms would be observed in the future. An obvious reduction in NO_X emissions for 2035 relative to 2015 would occur, largely in the transport sector, resulting from the high penetration of electric vehicles. Different from Beijing, although they are both municipalities, Tianjin would still be in rapid industrialization, and its power sector would remain dominated by its existing coal-fired power plants. Under the EGA scenario, the industrial sector and power sector in Tianjin would contribute the most of the SO_2, primary $PM_{2.5}$, and CO_2 emission reductions for 2035 relative to 2015. The largest NO_X emission reductions would occur in the transport sector in Tianjin.

Air Quality

The Current Policies scenario would achieve continued improvements in air quality, but the EGA scenario would fare much better in attaining the PRC's ambient air quality target of 35 µg/m³. In their annual mean $PM_{2.5}$ concentration by 2035, the following projected declines by region would result from the co-benefits of energy intensity and structural-change measures along with modest incremental end-of-pipe air pollution controls. The decline in Beijing would be by 55% to 36 µg/m³, in Tianjin by 36% to 45 µg/m³, in Hebei by 50% to 38 µg/m³, and in Shandong by 47% to 41 µg/m³ from 2015 to 2035. Still, these reductions relative to the 2015 levels would fall short of the national air quality goals.

In the EGA scenario over the same period, the greater efforts particularly in energy efficiency and fuel switching would substantially reduce air pollutant emissions. They would bring about the following declines in the annual mean $PM_{2.5}$ concentrations: in Beijing by 71% to 24 µg/m³, in Tianjin by 51% to 34 µg/m³, in Hebei by 62% to 29 µg/m³, and in Shandong by 59% to 31 µg/m³, respectively (Figure 13).

The annual mean $PM_{2.5}$ concentration in Beijing would decline to around 35 µg/m³ by 2025, and that in Hebei and Shandong would decline to around 35 µg/m³ by 2030.

All of these declines in the annual mean $PM_{2.5}$ concentrations would be consistent with the ambient air quality targets.

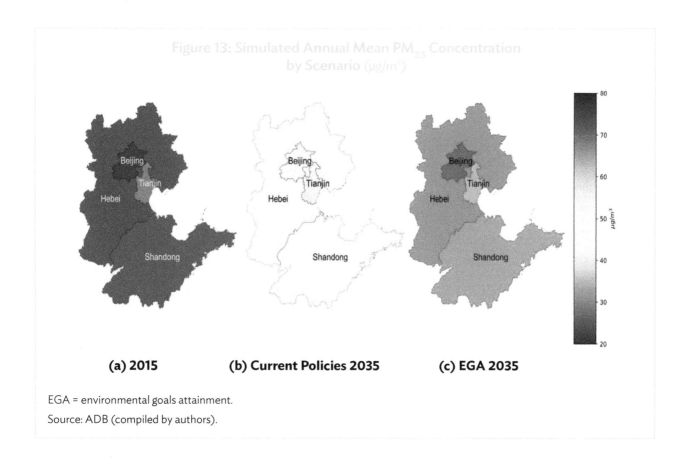

Figure 13: Simulated Annual Mean PM₂.₅ Concentration by Scenario (μg/m³)

(a) 2015   (b) Current Policies 2035   (c) EGA 2035

EGA = environmental goals attainment.

Source: ADB (compiled by authors).

Relative Impacts of the Two Strategies

Figure 14 shows how the incremental introduction of air pollution reduction policies in the Current Policies and EGA scenarios would translate into regionwide $PM_{2.5}$ concentration reductions. End-of-pipe control measures would deliver 28%–40% of the total reduction across provinces in 2035; on the other hand, 60%–72% of the reduction would be contributed by energy efficiency and fuel-switching measures that further replace coal with renewable energy and imported electricity. For the four provinces in BTH and Shandong, this clearly shows the crucial importance of energy efficiency and fuel-switching measures in improving their air quality.

The relative contribution of energy efficiency and of fuel switching in achieving the $PM_{2.5}$ target would vary by municipality/province: 61% in Beijing, 71% in Tianjin, 72% in Hebei, and 60% in Shandong.

An important distinction between this most recent study and other studies recently undertaken in the PRC is the difference in emphasis given to measures targeting energy demand and composition and that given to measures targeting end-of-pipe controls.

Prior studies have shown that between 2013 and 2017, the reductions in ambient $PM_{2.5}$ concentrations resulted mainly from end-of-pipe emissions control. The results

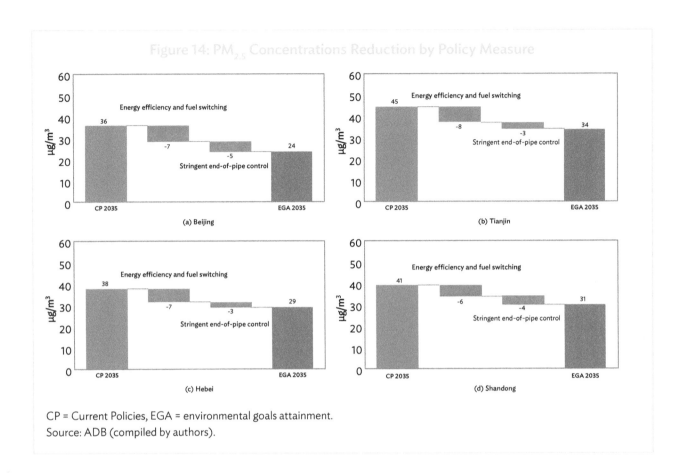

Figure 14: PM$_{2.5}$ Concentrations Reduction by Policy Measure

CP = Current Policies, EGA = environmental goals attainment.
Source: ADB (compiled by authors).

of this more recent integrated study clearly indicate that for improving air quality while at the same time addressing climate change mitigation, energy efficiency and fuel-switching measures are needed to attain air quality goals than just by end-of-pipe emissions control.

From both the cost standpoint and the sheer scale of control required within the 15-year horizon until 2035, it is clear that the study's integrated strategy would be much more effective and far less challenging than sole reliance on end-of-pipe controls to attain the targeted air-quality improvements.

Gross Domestic Product

The most important advantage of the study's framework is that it can estimate the economic impacts of policy measures to reduce air pollution levels. The economic impact was estimated by using the value of GDP loss in the EGA scenario relative to the Current Policies scenario.

In the Current Policies scenario, a prespecified GDP growth path was produced at first by taking into account the endogenous factors affecting growth, i.e., labor productivity and capital accumulation, and then by adding the additional policy constraints defined in the EGA scenario that would lead to the change in economic structure, in energy efficiency improvement, and in the need for fuel switching. Changes in GDP would consequently be captured, but this framework under the Current Policies scenario has the limitation of being incapable of capturing the costs of end-of-pipe control technologies.

By undertaking the Current Policies scenario and EGA scenario in tandem, the study has projected continued economic growth in BTH and Shandong through 2035.

In the Current Policies scenario, these regions would achieve annual average GDP growths from 2020 to 2035 as follows: 4.4% in Beijing, 4.4% in Tianjin, 4.7% in Hebei, and 4.8% in Shandong.

In the EGA scenario, these regions would each experience a deceleration of GDP growth rates by 2035: Beijing by 1.7 percentage points, Tianjin by 2.3 percentage points, Hebei by 1.4 percentage points, and Shandong by 0.8 percentage points (Figure 15). However, it is important to emphasize that these are relatively modest declines when viewed against the estimated health benefits that are discussed in detail.

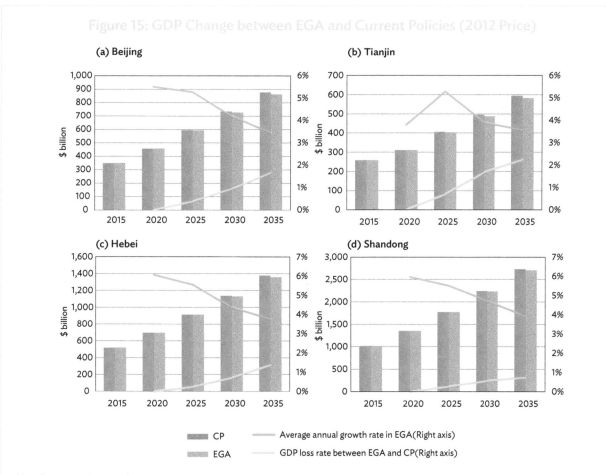

Figure 15: GDP Change between EGA and Current Policies (2012 Price)

CP = Current Policies, EGA = environmental goals attainment, GDP = gross domestic product.
Source: ADB (compiled by authors).

CO$_2$ Price

C-REM projects the CO$_2$ price associated with province-specific measures to achieve structural change and energy efficiency and fuel switching. To be able to show the marginal cost of abatement given each region's level of policy ambition, however, these measures must be modeled with different caps across provinces. This is because CO$_2$ prices would be higher in places with a greater preexisting decarbonization effort and that are implementing more stringent policies.

As shown in Figure 16, the CO$_2$ price required to achieve the reduction targets would vary by province, with prices generally higher in Beijing and Tianjin, which are less CO$_2$-intensive compared to Hebei and Shandong. This difference is largely accounted for by CO$_2$ reduction measures previously undertaken in Beijing and Tianjin that had reduced the remaining reduction opportunities at lower marginal cost.

In contrast, Shandong Province would retain the highest coal consumption in the PRC, so it would have abundant opportunities to reduce its use of coal at low marginal cost.

Under the Current Policies scenario relative to the EGA scenario, municipalities and provinces with less stringent emission reduction targets would have lower CO$_2$ prices.

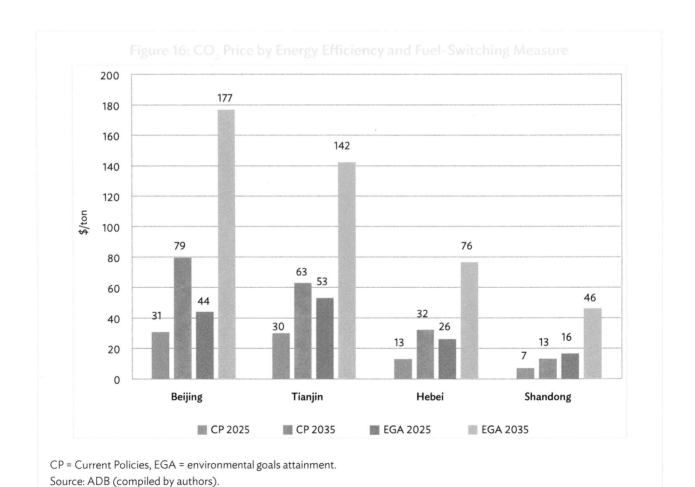

Figure 16: CO$_2$ Price by Energy Efficiency and Fuel-Switching Measure

CP = Current Policies, EGA = environmental goals attainment.
Source: ADB (compiled by authors).

Climate and Health Benefits

Using the REACH Framework, the study allowed the estimation of the health and climate benefits derived from improved air quality and reduced CO_2 emissions in both the Current Policies and EGA scenarios. Those benefits, the value of which is called the social cost of carbon (SCC), are estimated in terms of avoided damages caused by climate change.

The SCC reflects the cost of changes in net agricultural productivity, the cost of avoided climate-induced damage in human health, the cost of avoided property damages from increased flood risk, and so on. Based on prior literature, this particular study adopted an SCC value of $55/ton in 2035 (based on 2007 price).

Using this estimate, the total value of the avoided climate damage in the four regions by 2035 was projected to reach $34.8 billion, with Beijing contributing $1.5 billion; Tianjin, $2.8 billion; Hebei, $13.6 billion; and Shandong, $16.9 billion. This value was arrived at by multiplying the SCC with the incremental CO_2 emission reductions achieved by improved energy efficiency and fuel switching in the EGA scenario (Figure 17).

The air quality benefits consist of direct benefits (i.e., avoided direct labor and leisure loss from increased morbidities and mortalities) and indirect benefits (i.e., avoided efficiency loss from the interrupted optimal resource allocation).

The REACH framework computes the monetized benefits associated with avoided mortalities as a component of the direct benefits. Using its health effects module, REACH then computes the number of mortalities avoided due to incremental air-quality improvements in the EGA scenario compared to those in the Current Policies scenario. For consistency, the valuations focus only on the effect of the energy efficiency and fuel-switching measures.

To put a value to these avoided mortalities, the common practice of applying a value of a statistical life (VSL) is adopted. For health-related studies in the PRC, the VSL ranges from about $0.17 million to $7.75 million. In the study, both a low value of VSL ($0.42 million) and a relatively high value ($3.7 million) were applied.

Using these parameters, the health benefits due to avoided mortalities and morbidities in the four regions were estimated at between $16 billion and $142 billion. Using the lower (higher) bound VSL estimates, these benefits would be $4 (36) billion in Beijing, $2 (15) billion in Tianjin, $7 (65) billion in Hebei, and $3 (26) billion in Shandong.

The indirect air quality-related health effect measured by benefit on the domestic economy was projected to total $11 billion, with $5.1 billion contributed by Beijing, $1.7 billion by Tianjin, $3 billion by Hebei, and $0.9 billion by Shandong.

Comparing these benefits against the economic cost (measured by the GDP losses due to the energy efficiency and fuel-switching measures), the air quality co-benefits alone would already offset the costs of the energy efficiency and fuel-switching measures either partially or fully.

In the EGA scenario relative to the Current Policies scenario, the economic cost in 2035 of air quality co-benefits alone would be approximately $18 billion in Beijing, $16 billion in Tianjin, $22 billion in Hebei, and $11 billion in Shandong.

Using the higher value of the VSL ($3.7 million), the climate and health benefits therefore would far outweigh the economic cost. Even with the lower bound of the VSL, the value of climate and health co-benefits would already cover 61%, 39%, 107%, and 190% of the mitigation costs in Beijing, Tianjin, Hebei, and Shandong, respectively.

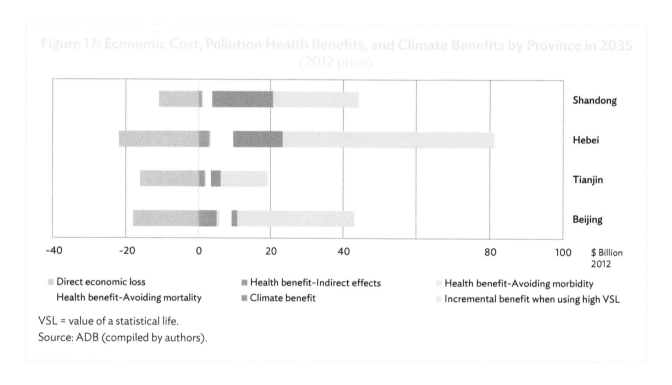

Figure 17: Economic Cost, Pollution Health Benefits, and Climate Benefits by Province in 2035 (2012 price)

VSL = value of a statistical life.
Source: ADB (compiled by authors).

Investment

Investment in the C-REM is represented by a sector that produces an investment that does well in using various inputs from different sectors. The output should be equivalent to the sum of the investments across all sectors and this total should be comparable across scenarios.

In the C-REM scenario, new investment capital is treated as savings and is determined by the total national income and the savings rate. The evolution of capital over time in the C-REM includes both old capital carried over from the previous period and new capital from investment made in the current period.

New capital investment is a key driver of GDP growth. To estimate this investment and its distribution by sector in the two policy scenarios, the cumulative changes in incremental capital required between 2020 and 2035 need to be reported.

In the EGA scenario, the regions were projected to require a total additional capital investment of $573 billion—$118 billion in Beijing, $77 billion in Tianjin, $125 billion in Hebei, and $253 billion in Shandong. Stringent measures of energy efficiency and fuel switching that put a price tag to CO_2 emissions would induce substitution away from CO_2-intensive activities. Most of the investment flows into the services sector and the other non-energy-intensive industries (Figure 18). The services sector receives 69% of the incremental capital in Beijing, 80% in Tianjin, 45% in Hebei, and 53% in Shandong, while other non-energy-intensive industries receive 20% in Beijing, 13% in Tianjin, 25% in Hebei, and 30% in Shandong. Beyond these two sectors, public transport, chemicals (including refined oil), and power generation (excluding coal-fired power plants) are also projected to experience large increases in investment in the EGA scenario. It should be noted that this does not reflect additional investment into end-of-pipe control equipment.

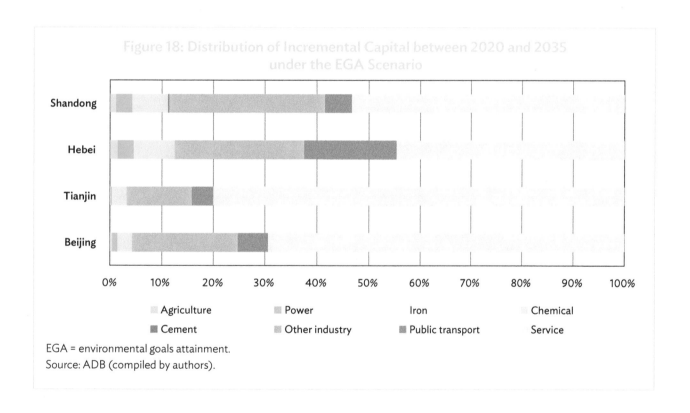

Figure 18: Distribution of Incremental Capital between 2020 and 2035 under the EGA Scenario

EGA = environmental goals attainment.
Source: ADB (compiled by authors).

6 Conclusions and Policy Recommendations

Conclusions

In the study, the status of CO_2 emissions and air pollution in BTH and Shandong were examined. After that, the major climate change and air pollution mitigation policies were simulated with a focus on a comparison among two scenarios that represent increasing degrees of policy effort.

The Current Policies and EGA scenarios were selected from a set of six that explored the implications of various policy measure combinations in two categories. The first category is energy efficiency and fuel switching, and the second category is end-of-pipe air pollution controls. All scenarios were simulated using the REACH integrated assessment modeling framework.

The EGA scenario was the only one that consistently achieved the PRC's ambient air quality targets. Very importantly, this scenario also demonstrated how energy efficiency and fuel switching could contribute to air-quality improvement.

For the period from 2015 to 2035, the EGA scenario projected a decline in annual average ambient $PM^{2.5}$ concentration of 71% to 24 µg/m^3 in Beijing, of 51% to 34 µg/m^3 in Tianjin, of 62% to 29 µg/m^3 in Hebei, and of 59% to 31 µg/m^3 in Shandong.

CO_2 emissions from the region are projected to decrease from 1.8 billion tons/year on average over the period 2015–2020 to 1.72 billion tons/year in 2025 and to 1.25 billion tons/year in 2035.

Compared to prior studies, the approach of the study had several distinctive features. First, it started with existing policies and identified a broader set of scenarios that vary effort on several dimensions. This process made it possible to develop packages of policies that the experts deemed sufficiently aggressive but feasible enough to implement. This analysis thereafter examined the expected impacts of these pathways and quantified a subset of the health and climate benefits as well as the economic costs that each path implies. The resulting suite of scenarios thus provided a strong basis for informed policy decisions.

By examining air quality, climate, and health outcomes within a single modeling framework, the study went far beyond most analytical tools for assessing policy impact. It clearly demonstrated the degree to which measures typically focused on reducing energy intensity or CO_2 emissions can contribute to addressing persistent air quality challenges in the PRC's major industrial regions.

Finally, the modeling framework of the study made it possible to show the impact of national policies on the provincial level by translating them into targets that take into account the linkages between the provincial economies.

The study was subject to several uncertainties and limitations. First, predicting the $PM_{2.5}$ concentrations remains a challenge because of high uncertainties in the air quality modeling. Second, by considering only the energy-related emissions reductions, it is possible that the study's projections might have somehow underestimated the potential of air-quality improvements. Third, a deeper look into the impact of various meteorological conditions on $PM_{2.5}$ responses to precursors is needed.

The remainder of this section is devoted to policy recommendations that emerged from the modeling analysis. They were arrived at in close consultation with the experts in the PRC who helped inform and develop the scenario designs for the study. This innovative consultative process gives insights not only into the anticipated impacts of various policies, but has informed a set of recommendations that support both climate change and air quality goals that are grounded in expert assessments of what measures are acceptable to stakeholders and thus can be immediately and confidently implemented on the ground.

Policy Recommendations

Based on the major findings, the study recommends the immediate enactment of seven major national policies by the PRC to achieve the climate and air quality goals for BTH and Shandong. These policies were arrived at in close consultation with experts in the PRC who helped inform, develop, and validate the scenario designs for the study.

Set Absolute Caps for CO_2 Emissions and $PM_{2.5}$ Concentrations for the Fourteenth Five-Year Plan and Beyond, and Extend Their Realization to at Least the Year 2035

In line with the Twelfth and Thirteenth Five-Year Plans, the provinces in the PRC have pursued their respective energy and CO_2 intensity targets. These targets were broadly accepted and vigorously pursued, but the provinces still must put the PRC on a path to more solid footing to attain the 2°C target and achieve their own part of the National Ambient Air Quality Standards.

For this reason, municipalities and provinces in BTH and Shandong need to set an absolute cap for CO_2 emissions and ambient $PM_{2.5}$ pollution in their Five-Year Plan through 2035.

Reduction of CO$_2$ Emissions

The recommended absolute cap targets for their reductions in CO$_2$ emissions are as follows:

(i)   **From 2015 to 2025:** 17% reduction in Beijing, 17% reduction in Tianjin, 6% reduction in Hebei, and 4% reduction in Shandong
(ii) **From 2015 to 2030:** 29% reduction in Beijing, 29% reduction in Tianjin, 16% reduction in Hebei, and 14% reduction in Shandong.

These CO$_2$ emissions reduction targets are based on the EGA scenario, and they are much more aggressive than the country's existing Nationally Determined Contribution under the Paris Agreement.

Reduction of ambient PM$_{2.5}$ pollution

To address ambient PM$^{2.5}$ pollution, the four municipalities and provinces have to make binding targets to limit their PM$^{2.5}$ concentrations as follows:

(i)   **By 2025:** 35 µg/m^3 in Beijing, 45 µg/m^3 in Tianjin, 40 µg/m^3 in Hebei, and 40 µg/m^3 in Shandong for 2025
(ii) **By 2030:** 30 µg/m^3 in Beijing, 40 µg/m^3 in Tianjin, 35 µg/m^3 in Hebei, and 35 µg/m^3 in Shandong.

The EGA scenario provides guidance on direct emissions levels that are consistent with achieving these tougher limits.

Accelerate Efforts to Reduce Coal Consumption through 2035

Coal combustion is a major source of air pollution and CO$_2$ emissions in BTH and Shandong. The study's analysis indicates that reducing coal consumption is crucial to achieving both the PRC's climate and air quality goals. The phasing out of coal should, therefore, be aggressively implemented by BTH and Shandong as part of their Fourteenth and Fifteenth Five-Year Plans.

Implement Renewable Portfolio Standards for Wind and Solar Power

The costs of generating electricity from wind and solar power are rapidly decreasing, and by 2025 these sources are expected to be priced at a competitive level with coal electricity in the region. The policies for renewable portfolio standards and green electricity dispatch schemes should therefore be aggressively implemented to help create greater demand for wind and solar power as electricity sources.

Continually Expand the Generation of Renewable Energy to Ensure the Stability and Reliability of the Region's Electricity Supply

First, greater integration of the power grid in the region should be pursued by improving better grid interconnections so as to enlarge the balancing area and provide greater absorption of power supply instabilities. This integration will involve building new transmission lines and changing operating rules to allow for more flexible inter-provincial transfer of electricity.

Second, flexibly dispatchable power stations and storage should be added to balance instabilities when they arise. As the cost of energy storage drops, grid storage will likewise become a more important opportunity for the region. The region can more strongly tap this opportunity by formulating a clear policy on how energy storage facilities should be compensated.

Third, the development of demand response should be encouraged so as to incentivize reductions on shifts in load when needed; this will help balance supply and demand in the presence of high shares of renewables. BTH and Shandong can provide a robust policy framework for the demand response and pilot its implementation.

Fourth and finally, electricity market reforms should be accelerated to increase demand for renewable power, which has the advantage of lower marginal production cost compared to other sources. The responsiveness of alternative power sources likewise should be improved to balance renewable power sources when needed via the price signal. To achieve this balancing effort, an ancillary services market must be developed to encourage investments in the needed flexible generation capacity.

Introduce an Adequately High Carbon Price Either through the National CO_2 Emissions Trading Program or through a Carbon Tax

An adequately high CO_2 price is vital to mobilizing investment and setting the direction for a green, low-carbon transformation such as the one modeled in the EGA scenario. A national carbon market has been under development in the PRC for most of the last decade, and was officially launched in 2021.

The national emissions trading program offers the region a chance to use CO_2 pricing to advance its air quality and climate goals. The study recommends three major action steps on how the region can make the most out of this opportunity.

First, as the EGA scenario makes clear, CO_2 pricing will incentivize both economic structural change and energy efficiency and fuel switching as a means for substantially contributing to both air quality improvement and climate change mitigation. BTH and Shandong should actively participate in the program and play a leading role in its development, both at the provincial/municipal and regional levels.

Second, the governments of the four municipalities and provinces should set a tight emissions cap for entities that are participating in the national emissions trading program. The trading should be limited to the region alone to prevent the reductions achieved under the program from being used as credit by other participating entities outside the region. Limiting trading to the region would ensure that emitters will undertake reductions on their own and directly reap co-benefits consistent with their own air quality goals, rather than outsourcing or trading their reduction credits to or from distant provinces.

Finally, a scheme for pricing CO_2 in the rest of the economy, such as a carbon tax, should be considered to ensure that all sectors are provided incentives fairly and equitably for undertaking a green, low-carbon transition.

Increase the Stringency of End-of-Pipe SO_2, NO_x, and PM Emissions Standards in All Industrial Sectors

The ultra-low emissions standards for SO_2, NO_x, and PM that have been implemented in the power generation sector in recent years have contributed substantially to air-quality improvement. This trend will continue if similarly tough standards are likewise introduced for non-power industrial emissions sources.

Emission standards of air pollutants for iron and steel, cement, flat glass, refined aluminum, coking, brickmaking, lime-making, petroleum refining, sulfuric acid, nitric acid, and industrial boilers, etc., need to be tightened. For example, the study suggests that the recommended regional targets to reduce the ambient $PM_{2.5}$ pollution would be greatly supported by limiting average hourly concentrations of PM, SO_2, and NO_x emissions from the sintering process to no more than 10 mg/m^3, 35 mg/m^3, and 50 mg/m^3 in the iron and steel sector.

Beyond these targeted for the various industries, further contributions to air-quality improvement could be realized by policy measures to reduce emissions of residential non-point sources (e.g., district heating), emissions of fugitive dust emissions, and emissions from agriculture and other natural sources.

APPENDIX 1
Detailed Assumptions on Policy Scenarios

Measures			Current Policies	Environmental Goals Attainment
Energy efficiency and fuel switching	Upper limits on CO_2 emissions		4.5% of annual average CO_2 intensity reduction from 2015 to 2035 for each province CO_2 emissions in the region would peak in 2025.	6.5% of annual average CO_2 intensity reduction from 2015 to 2035 for each province CO_2 emissions in the region would peak by 2025 and would decrease to 1.27 billion tons by 2035.
	Renewable energy	Wind power	**Beijing**: The installed capacity would increase to 2.5 GW by 2035.	**Beijing**: The installed capacity would increase to 3.5 GW by 2035.
			Tianjin: The installed capacity would increase to 1.9 GW by 2035.	**Tianjin**: The installed capacity would increase to 2.5 GW by 2035.
			Hebei: The installed capacity would increase to 30 GW by 2035.	**Hebei**: The installed capacity would increase to 97 GW by 2035.
			Shandong: The installed capacity would increase to 25 GW by 2035.	**Shandong**: The installed capacity would increase to 47 GW by 2035.
		Solar power	**Beijing**: The installed capacity would increase to 7.2 GW by 2035.	**Beijing**: The installed capacity would increase to 12.5 GW by 2035.
			Tianjin: The installed capacity would increase to 3 GW by 2035.	**Tianjin**: The installed capacity would increase to 6 GW by 2035.
			Hebei: The installed capacity would increase to 36 GW by 2035.	**Hebei**: The installed capacity would increase to 79 GW by 2035.
			Shandong: The installed capacity would increase to 30 GW by 2035.	**Shandong**: The installed capacity would increase to 41 GW by 2035.
		Nuclear power	**Shandong:** The installed capacity would increase to 21 GW by 2035.	**Shandong:** The installed capacity would increase to 37 GW by 2035.
		Hydro power	**Nation**: Achieve the existing target of 340 GW in 2020 and increase it to 460 GW by 2035.	
			The resources of hydro power in Beijing, Tianjin, Hebei, and Shandong would remain very limited.	
		Subsidies	CNY0.01/kWh subsidy for wind and CNY0.08/kWh subsidy for solar in 2020, subsidies for wind and solar would be phased out before 2025.	
	Electric vehicles		The ownership of electric passenger vehicles would continue to grow. By 2035, the penetration rate of electric passenger vehicles in Tianjin, Hebei, and Shandong would reach 17.2%, and would reach 28% in Beijing.	The ownership of electric passenger vehicles would grow rapidly. By 2035, the penetration rate of electric passenger vehicles in Tianjin, Hebei, and Shandong would reach 32.2%, and in Beijing would reach 60%.

continued on next page

Table continued

Measures		Current Policies	Environmental Goals Attainment
End-of-pipe control	Ultra-low emissions technologies in the power sector	The desulfurization and denitrification efficiency would increase after 2020—the desulfurization efficiency to 98.8% and the denitrification efficiency to 91.0% by 2035. The application rate of ESP-FF and FF would increase to about 55% in 2035.	The desulfurization and denitrification efficiency would increase after the 2020—the desulfurization efficiency to 99.0% and the denitrification efficiency of 91.5% to 2035. The application rate of ESP-FF and FF would increase to about 65% in 2035.
	Pollution control and upgrading in key industries* **Cement:** For desulfurization, the application rate of Flue Gas Desulfurization (FGD) would increase to about 80% by 2035. For denitrification, the application rate of Low NOx burning (LNB)+SCR+SNCR would increase to about 75% by 2035. For dust removal, the application rate of ESP-FF would increase to 20% by 2035. **Flat glass:** For desulfurization, the application rate of High-efficiency flue gas desulfurization (HEFGD) would increase to about 55% by 2035. For denitrification, the application rate of SCR would increase to about 95% by 2035. For dust removal, the application rate of FF and ESP+WESP would increase to 80% by 2035.	**Iron and steel:** For desulfurization, the application rate of denitrification facilities would be 100% by 2030, and that of highly effective SCR denitrification technology would increase to 40% by 2035. For dust removal, the application rate of ESP-FF would increase to 20% by 2035. **Cement:** For desulfurization, the application rate of FGD would increase to about 95% by 2035. For denitrification, the application rate of LNB+SCR+SNCR would increase to about 95% by 2035. For dust removal, the application rate of ESP-FF would increase to 35% by 2035. **Flat glass:** For desulfurization, the application rate of HEFGD would increase to about 80% by 2035. For denitrification, the application rate of SCR would increase to about 100% by 2035. For dust removal, the application rate of FF and ESP+WESP would increase to 95% by 2035.	**Iron and steel:** For desulfurization, the application rate of denitrification facilities would be 100% by 2030, and that of highly effective SCR denitrification technology would increase to 75% by 2035. For dust removal, the application rate of ESP-FF would increase to 30% by 2035.
	Ultra-low emission transformation Industrial coal-fired boilers	For desulfurization, the application rate of HEFGD would increase to about 95% by 2035. For denitrification, the application rate of highly efficient LNB+SCR would increase to about 45% by 2035. For dust removal, the application rate of ESP+WESP would increase to 10% by 2035.	For desulfurization, the application rate of HEFGD would increase to about 100% by 2035. For denitrification, the application rate of highly efficient LNB+SCR would increase to about 65% by 2035. For dust removal, the application rate of ESP+WESP would increase to 20% by 2035.
	Promoting advanced stoves in the building sector	65% bulk coal used for heating or cooking would be substituted by the briquette or advanced clean stoves by 2035.	All bulk coal used for heating or cooking would be substituted by the briquette or advanced clean stoves by 2030.
	Fuel quality improvement and emission standards upgrade of vehicles	The strictest implementation of emission standards is assumed according to the currently specified schedules in both two scenarios.	

Notes: * The assumptions for iron and steel, cement, and flat glass industry are provided here. More details on other industries are provided in Wang (2019), e.g., on refined aluminum, coking, brickmaking, lime-making, petroleum refining, sulfuric acid, and nitric acid.

Source: Authors

APPENDIX 2
Detailed Description of the REACH Framework

The Regional Emissions Air-Quality Climate and Health (REACH) framework was first developed in Li et al. (2018). At both a provincial and a national scale in the People's Republic of China (PRC), the study usedthe China Regional Energy Model (C-REM), the REAS/MEIC emissions inventory, the GEOS-Chem atmospheric chemistry model, and BenMAP to assess air quality co-benefits of carbon pricing. As described in Zhang et al. (2017), this framework was expanded in 2015 to include a health service sector and leisure-consumption substitution to support estimates of the indirect economic burden of air pollution in the PRC.

China Regional Energy Model

As the core model of the integrated assessment framework of REACH, C-REM is a global multiregional, multi-sector, dynamic recursive computable general equilibrium model. This model has been used in several previous analyses to evaluate the PRC's energy and climate policy.

Model Regions and Sectors

The C-REM focuses on the impact of industrial and climate policies on various regions in the PRC. Considering the data availability, 30 provinces were described in C-REM as shown in Table A2.1.

Table A2.1: Regions Represented in the C-REM

Provinces	Label	Provinces	Label
Beijing	BJ	Henan	HA
Tianjin	TJ	Hubei	HB
Hebei	HE	Hunan	HN
Shanxi	SX	Guangdong	GD
Inner Mongolia	NM	Guangxi	GX
Liaoning	LN	Hainan	HI
Jilin	JL	Chongqing	CQ
Heilongjiang	HL	Sichuan	SC

continued on next page

Table A2.1 *continued*

Provinces	Label	Provinces	Label
Shanghai	SH	Guizhou	GZ
Jiangsu	JS	Yunnan	YN
Zhejiang	ZJ	Shaanxi	SN
Anhui	AH	Gansu	GS
Fujian	FJ	Qinghai	QH
Jiangxi	JX	Ningxia	NX
Shandong	SD	Xinjiang	XJ

C-REM = China Regional Energy Model.
Source: Authors.

The C-REM aggregates the economic activities of each province into 21 sectors. Shown in Table A2.2 are the C-REM sectorial aggregation and description. It is worth noting that in earlier versions of the C-REM, energy-intensive sectors such as iron and steel, nonmetallic products, nonferrous metal products, and the chemical sector were aggregated into a single sector, whereas in the recently updated third version of C-REM, the number of production sectors has been extended from 13 to 21 based on multiregional input–output in 2012.

Moreover, to characterize regional electricity production, consumption, and trade, the C-REM model considers various power generation technologies, including coal, gas, oil, wind, solar, nuclear, hydro, and biomass power generation technologies.

Table A2.2: Descriptions of Production Sectors in the C-REM

Sector Type	Sector	Label	Description
Agricultural	Agricultural	AGR	Food and non-food crops produced on managed cropland, managed forestland and logging activities, animal husbandry and animal products
Energy	Coal	COL	Mining and agglomeration of coal and coking
	Crude oil	OIL	Extraction of petroleum
	Natural gas	GAS	Extraction and supply of natural gas
	Refined oil	ROI	Refined oil and petrol chemistry products
	Power	ELE	Electricity and heat generation, transmission, and distribution
Energy-intensive	Chemical	CRP	Basic chemicals, other chemical products, rubber, and plastics
	Iron and steel	ISM	Manufacture and casting of iron and steel, copper, aluminum, zinc, lead, gold, and silver
	Nonmetallic	NMM	Manufacture of cement, plaster, lime, gravel, and concrete

continued on next page

Table A2.2 *continued*

Sector Type	Sector	Label	Description
Other industry	Water	WAT	Production and processing of water
	Mining	MIN	Mining of metal ores, uranium, gems, and other mining/quarrying
	Transport equipment	TEQ	Transportation equipment
	Food and tobacco	FOD	Manufacture of food products and tobacco
	Textile	TWL	Manufacture of textiles
	Electronic equipment	ELQ	Electronic equipment
	Other industries	OTH	Other industries
	Other machinery	OME	Other machinery
Construction	Construction	CNS	Construction of houses, factories, offices, and roads
Transport	Transport	TRA	Pipeline transport, and water, air, and land transport (passenger and freight)
Services	Commercial and public services	SER	Communication, finance, public services
	Dwelling	DWE	Dwelling

C-REM = China Regional Energy Model.
Source: Authors.

Data Sources

As a general equilibrium model for studying issues in the provinces of the PRC, the database of C-REM includes economic data such as production, consumption, and regional trade in various sectors, and integrates physical data of production, consumption, and trade corresponding to economic quantities.

The provincial economic data are mainly based on the 2012 provincial input–output table. The provincial energy data are derived from the 2012 regional energy balance table (including energy production, consumption, processing conversion, and import and export data of each province) and from the China Electric Power Yearbook (for power generation data from various provinces including thermal, nuclear, and renewable energy generation). For other regions outside the PRC, C-REM uses data from GTAP9 of the Global Trade Analysis Project.

The input–output data that C-REM needs to read are first preprocessed, including combining error residuals, adjusting capital inputs, and correcting factor inputs. Then, due to the large differences in the sector classification of energy data and economic data, sector aggregation processing is carried out for different energy varieties. The raw coal, washed coal, briquette, and coke are aggregated into coal; the gasoline, diesel, kerosene, and fuel oils are aggregated into refined oils; and heat and electricity are aggregated into one category, as shown in Table A2.3.

Table A2.3: Aggregation of Energy Types in the C-REM

Aggregated	Label	Energy Types in Energy Database
Coal	COAL	Raw coal
		Cleaned coal
		Other cleaned coal
		Briquette
		Coke
		Coke oven gas
		Other gas
		Other coke
Crude oil	OIL	Crude oil
Refined oil	ROIL	Gasoline
		Kerosene
		Diesel
		Fuel oil
		Liquefied petroleum gas
		Refinery dry gas
		Other petroleum products
Gas	GAS	Natural gas
		Liquefied natural gas
Electricity and heat	ELEC	Heat
		Electricity

C-REM = China Regional Energy Model.
Source: Authors.

Model Structure and Key Parameters

In general equilibrium theory, markets are divided into product markets and factor markets, and economic entities are divided into producers and consumers.

Producers purchase labor, capital, and other resources from the factor market and other products and services from the product market. They are used as intermediate inputs for producing specific products or services.

Consumers sell labor, capital, and other resource endowments in the factor market to obtain income, and purchase products and services from the product market for their consumption or utility.

Producers determine their output and demand for factors of production at a given price to maximize profits; consumers at the same time provide factors of production or purchase products at a given price given particular budgets to maximize their utility.

The economic system reaches a general equilibrium when each market reaches a balance between supply and demand at the corresponding price.

The production function in C-REM can be described by the nesting structure of the constant elasticity of substitution (CES) function. The inputs under the same layer of CES nesting structure can replace one another in the production activities with particular degrees of substitution elasticity.

Equation B-1 is a typical dual-input CES production function that takes capital (K) and labor (L) inputs as examples:

$$X_{KL} = [\theta K^Y + (1 - \theta)L^Y]^{\frac{1}{Y}} \tag{B-1}$$

where θ is the ratio of capital to total input ($1 - \theta$ is the ratio of labor to total input), Y is related to the elasticity of substitution σ, $\sigma = 1/(1 - Y)$, and X_{KL} is the corresponding output. When there are more than two kinds of inputs for production activities, it is sufficient to adjust the input proportion coefficients accordingly so that their sum is equal to 1, but this form is only effective when the inputs have the same elasticity of substitution.

When the substitution elasticity is different, a new level of CES production function is introduced. In Equation B-2, e.g., the above-mentioned capital–labor service X_{KL} and energy E are used as inputs to produce a new product Y, as follows:

$$Y = [\theta_E(E)^{Y_E} + (1 - \theta_E)(X_{KL})^{Y_E}]^{1/Y_E} \tag{B-2}$$

By adding a new substitution level, a new substitution elasticity Y_E different from the value of Y can be selected. Therefore, the production function can be expressed in various ways through the multilayered CES structure.

For example, the production function of fossil energy production department ff in region r in C-REM can be expressed as Equation B-3:

$$Y_{ff,r} = \left[\alpha_{ff,r} R_{ff,r}^{\rho_{r\_oth}} \right. \tag{B-3}$$
$$\left. + \left(1 - \alpha_{ff,r}\right) \cdot min\left(M_{1,ff,r}, \dots, M_{i,ff,r}, E_{ff,r}, V_{ff,r}\right)^{\rho_{r\_oth}}\right]^{1/\rho_{r\_oth}}$$

In Equation B-3, $\alpha_{ff,r}$ is the ratio of the corresponding energy resources $R_{ff,r}$ to the total input; ρ_{r_oth} is related to the substitution elasticity σ_{r_oth}; $\sigma_{r\_oth} = 1/\rho_{r\_oth}$, $M_{i,ff,r}$ is the intermediate input; $E_{ff,r}$ is the energy input; and $V_{ff,r}$ is the value-added bundle including capital and labor inputs.

The production function of $V_{ff,r}$ satisfies the form of the Cobb–Douglas production function (such as Equation B-4, where β_1 is the ratio of labor input to the cluster input), and $Y_{ff,r}$ is the output of the regional fossil energy production sector.

$$V_{ff,r} = L_{ff,r}^{\beta_1} \cdot K_{ff,r}^{1-\beta_1} \tag{B-4}$$

Similar to the production function structure, C-REM also applies the nested CES function to describe the consumption behavior of governments and residents, as shown in Figure A2.1. The top level of the consumption function structure is the substitution relationship between consumption and savings.

At the same time, to track the impact of energy consumption in the consumer sector under the policy, C-REM divides goods and services into non-energy and energy categories; the latter category includes coal, refined oil, natural gas, and others.

When considering the health impact of increasing the air pollution medical service production sector, it must be noted that the nested structure of the consumption function needs to be improved to reflect the impact of air pollution on residents' consumption.

Figure A2.1: The Structure for Residential Consumption

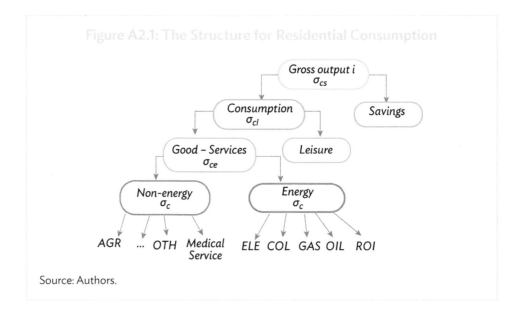

Source: Authors.

In the CES function of the main production and consumption in the C-REM, the value of substitution elasticity is determined based on the MIT EPPA model. To make the model more suitable for the characteristics of the PRC's energy production and consumption, the elasticity of substitution may need adjustment in some CES functions; as, e.g., by adjusting the elasticity of substitution between coal, oil, and gas inputs in the production sector. Table A2.4 shows the substitution elasticity values.

Table A2.4: Substitution Elasticities in the Production and Consumption Sector

Parameter	Substitution Margin	Value
σ_{en}	Energy (excluding electricity)	1
σ_{enoe}	Energy—electricity	0.5
σ_{eva}	Energy/electricity—value-added	0.5
σ_{kl}	Capital—labor	1
σ_{klem}	Capital/labor/energy—materials	0.5
σ_{cong}	Coal/oil—natural gas in ELE	2
σ_{co}	Coal—oil in ELE	2
σ_{hn}	Resource—capital/labor/energy/materials/hydro/nuclear/wind ELE	0.01
σ_{ws}	Resource—capital/labor/energy/materials in wind/solar ELE	0.2
σ_{rcol}	Capital/labor/materials—resources in COL	0.7
σ_{roil}	Capital/labor/materials—resources in OIL	0.6
σ_{rgas}	Capital/labor/materials—resources in GAS	1.2
σ_{genoe}	Materials—energy in government demand	0.5
σ_{gen}	Energy in government demand	0.1
σ_{gnoe}	Materials in government demand	1
σ_{ct}	Transportation—non-transport in private consumption	0.5
σ_{ce}	Energy—non-energy	0.3
σ_{c}	Non-energy in private consumption	0.3
σ_{ef}	Energy in private consumption	0.5

Source: Authors.

In C-REM, the economic growth of each region is driven mainly by the growth of labor and capital supply, while the labor supply of each region is mainly determined by population and labor productivity, as shown in Equation B-5:

$$L_{r,t} = L_{r,t-1} \cdot (1 + n_{r,t})^u \cdot (1 + g_{r,t})^u \qquad \text{(B-5)}$$

In the equation, u is a unit time period, which is 5 years in C-REM, $L_{r,t}$ is the labor supply of region r at time t, and $n_{r,t}$ and $g_{r,t}$ stand for the average annual growth rates of population and labor productivity, respectively.

Multi-Pollutant Emissions Inventory

In the study, the air-quality impact assessment was conducted using the multi-pollutant emissions inventory[1] and the air quality model.

The multi-pollutant emissions inventory is an accounting model to calculate the number of pollutants discharged into the atmosphere. It is a useful tool for the development of an effective air-quality management process. In particular, the Chinese Emission Inventory, which was developed and updated by Tsinghua University, was used in the study to calculate the number of pollutants in BTH and Shandong.

The "emission factor"[2] method that was used in the study has three classes of parameters: (i) the level of activities (e.g., fuel consumption, the industrial product yield, solvent usage, etc.), (ii) the technology-based uncontrolled emission factors, and (iii) the penetrations of control technologies. The detailed methodology and data sources can be found in previous papers.

The base year of the multi-pollutant emissions inventory is 2015. The inventory is composed of both the following pollutants, like SO_2, NO_x, Non-Methane Volatile Organic Compounds (or NMVOC), and NH_3, and the primary particulate matters, like PM_{10}, $PM_{2.5}$, black carbon, and organic carbon that contribute to secondary $PM_{2.5}$ formation.

Primary emission source is divided into seven categories as shown in Table A2.5. According to the emission characteristics of the various pollutants, each emission source is further divided into four levels: sectors, fuel and/or products, combustion and/or technology, and control measures.

Both the emissions from anthropogenic sources and natural sources are considered. The natural sources emissions are taken from the MEGAN, which is listed under other emission sources.

---

[1]   An emission inventory is an inventory with details in the amounts and types of pollutants released into the environment (Source: UNEP. 2019. *Air Pollution in Asia and the Pacific: Science-Based Solutions.* https://www.ccacoalition.org/en/resources/air-pollution-asia-and-pacific-science-based-solutions-summary-full-report).

[2]   An emission factor is a unique value for scaling emissions to activity data in terms of a standard rate of emissions per unit of activity (Source: UNEP. 2019. *Air Pollution in Asia and the Pacific: Science-Based Solutions.* https://www.ccacoalition.org/en/resources/air-pollution-asia-and-pacific-science-based-solutions-summary-full-report).

Table A2.5: Primary Anthropogenic Emission Sources

Code	Sources	Description
1	Stationary combustion sources	Utilize the heat and water vapor from fuel combustion to provide energy for industrial production and daily life
2	Industrial sources	Air pollutants emission in industrial production process
3	Mobile sources	Air pollutants emission of transportation facilities and equipment
4	Solvent products usage	Volatile organic compounds emission in the process of organic solvents use
5	Fuel storage and transportation	Potential emission in fuel collection, storage, transportation, and sales process
6	Agricultural sources	Air pollutants emission in the agricultural production process
7	Other	Other emission sources

Source: Authors.

Air Quality Model

The air quality model is a mathematical model to describe the behavior of air pollutants in the atmosphere; it is based on the latest research results of atmospheric science. The model reproduces the physical and chemical processes of pollutants in the atmosphere to simulate the air quality based on the emission information of pollution sources and real-time meteorological data.

Over time since the 1960s, the air quality model has evolved into various designs. Community Multi-scale Air Quality, which is a numerical air quality model developed by the United States Environmental Protection Agency, is one of the most widely used ones.

To quickly describe the response of pollutant concentrations to emission changes in the multiple future scenarios, the study used the response surface model (RSM) to simulate $PM_{2.5}$ concentrations in 2025, 2030, and 2035 under different scenarios. The RSM is a reduced-form air quality prediction model using statistical correlation structures or polynomial functions to approximate model functions through the design of complex multidimensional experiments.

The RSM was initialized in 2006 by the United States Environmental Protection Agency, and has been continuously developed under the framework of the Air Benefit and Cost and Attainment Assessment System (ABaCAS, http://www.abacas-dss.com). The key parameters of RSM development have been tested and determined through computational experiments and validations, confirming the applicability of the RSM technique in the field of air quality modeling for $PM_{2.5}$ simulation.

An extended response surface model (ERSM) technique, proposed in 2011, was applied in a study of $PM_{2.5}$ concentration simulation in the PRC's Yangtze River Delta region. In the study, an updated ERSM (ERSMv2.0) developed in 2017 was used. For the new version, the ERSM method was improved by adding an estimation of indirect effects to represent the interaction among regions. The improvement has shown great advantages in dealing with the interaction of air-quality impact between adjacent regions.

A case study of the Beijing–Tianjin–Hebei region using the new version has already been carried out. January, April, July, and October were chosen as the typical months, with a simulation resolution of 27 km. The average of these four months was taken as the average annual concentration value, and the average of grid concentrations of all national control points in each province were taken as its average concentration. The typical weather year selected was the year 2017.

Health Effects Evaluation Module

The CREM-Health Effect Module (CREM-HE) is the extended health effect evaluation module in C-REM. It absorbs the latest epidemiologic study findings on the impact of air pollution on health and describes the exposure–response relationships of acute exposure health outcomes and chronic exposure deaths. This establishes the correspondence from the concentration to the incidence or death cases.

CREM-HE can dynamically depict the impact of increased morbidity and mortality on the supply of labor in factor markets caused by air pollution, and it can track the cumulative effects of long-term changes in the supply of labor in the economic system.

As an important part of the REACH model, the CREM-HE module not only can track the impact of air pollutants produced by economic systems on the public health level but can also describe the feedback effect of such changes in public health level on the economic system.

CREM-HE has the following three main characteristics:

First, unlike traditional research methods that only perform "slicing" analysis for specific years, CREM-HE can dynamically characterize the supply shock in factor market caused by the increase in morbidity and mortality due to air pollution; also, it can track the cumulative impact of economic systems on long-term changes in labor supply. This way, it more accurately describes the impact mechanism of long-term air pollution on economic systems.

Second, CREM-HE is based on the general equilibrium model theory and has the advantage of being able to capture the interaction between different sectors in the economic system. The market efficiency is highest in the case of optimal allocation of factors and commodities.

When the supply of factors is affected by external shocks, the optimal allocation of market resources is disrupted, and the market efficiency gets reduced. This part of the impact is called an indirect loss. CREM-HE extends the scope of air pollution impacts, including the direct and indirect losses, and the impact assessment is more scientific and comprehensive.

Third, CREM-HE pays attention to localization and differentiation of parameters in describing the exposure–response relationship. This model, on the one hand, draws on the latest epidemiology focusing on the PRC's analyses; on the other hand, it fully considers the differences between cases and deaths in different provinces and different age groups. This way it can more objectively reflect the impact of air pollution in the PRC at the provincial level.

Extended Health Service Sector

To cope with the health effects caused by air pollution, residents typically not only need to pay for rest and pay for the loss of salary but also need to pay more to purchase medical and health services. Therefore, the health effects of air pollution consist not only of loss of labor but also of loss in terms of medical expenses and leisure time. To analyze the above impact, the traditional social accounting matrix is improved and the production sector's loss of leisure time and need for health services are augmented to deal with the adverse effects of air pollution.

The portrayal of leisure time in the study thus reflects the welfare loss caused by the health impact, and the establishment of a health service department to deal with air pollution captures the economic investment corresponding to the residents' medical and health needs.

Figure A2.2 shows the expanded social accounting matrix. The production input of the air pollution health service sector includes labor and intermediary medical services. The sector produces health services to eliminate the adverse effects of air pollution.

For different health outcomes, the proportion of labor and medical services corresponding to production health services also varies greatly. For example, the proportion of medical expenses required for hospitalization for respiratory diseases and for cardiovascular and cerebrovascular diseases is much greater than that required for treatment at the emergency department.

At the same time, there are differences in the health effects experienced by different age groups. For example, when the elderly suffers from illness and succumbs to it, the cost of such health outcomes only includes medical expenses and leisure time loss but does not include labor loss.

As the concentration of pollutants increases, health service departments that respond to air pollution would need to invest more medical services and labor to produce health services, which would have a ripple effect on other sectors of the economic system. As the demand for health services increases, more labor and capital would be transferred from other sectors to this sector, which in turn would influence the prices and allocation of factors and commodities in the market.

The deterioration of air quality would lead to a decline in the social welfare of the economic system. On the one hand, it would result in the reduction of the labor force; on the other hand, the increase in demand for health services would drive

Figure A2.2: Expanded Social Accounting Matrix of CREM-HE

		Intermediate Input						Residents' Service		Final Use				Total Output
		1	2	…	j	…	n	Health Services	Labor/Leisure Option	Resident	Investment	Government	Export	
Production sectors	1													
	2													
	:													
	:													
	i													
	:													
	:													
	Medical service caused by pollution							Medical service			Health services	Health services		
	m													
Import	1													
	2													
	:													
	:													
	i													
	:													
	:													
	r													
Leisure time									Leisure time	Leisure time				
Value added	Labor							Labor	Labor					
	Capital													
	Resources													
	…													
Total Input														

CREM-HE = CREM-Health Effect Module.

Source: Zhang, X. 2016. Development and application of regional energy emission air-quality climate health model (REACH). Beijing: Tsinghua University.

the relocation of labor factors elsewhere in the market, and the efficiency of market resource allocation would consequently decline.

The nested consumption function after the health service sector for air pollution is added. Leisure time is added to characterize potential labor. This additional labor will not be supplied to the economic system as labor, but used to track how residents assess the value of leisure time. In CREM-HE, the quantitative assumption of leisure time follows the theory of marginal economy: residents believe that the value of labor time and leisure time is equal. Besides, to recover from the health damage caused by air pollution, residents need to consume air pollution health services in non-energy goods and services clusters. That is, when the concentration of air pollution increases, residents' consumption level of health services will increase accordingly.

Establishing Exposure–Response Relationships

In the CREM-HE, exposure–response functions are established and the relative risks (RR) are used to estimate the quantitative relationship between health outcomes and air pollution exposure. Equations B-6 and B-7 are used for acute exposure health effects.

$$Cases_{itr}^{AE} = [1 - 1/RR_i(C_{tr})] \cdot F_{itr} \cdot P_{tr} \qquad \text{(B-6)}$$

$$RR_i(C_{tr}) = e^{\beta_i C_{tr}} \qquad \text{(B-7)}$$

In the equations, $Cases_{itr}^{AE}$ represents the number of cases for acute exposure health outcome i in time t in region r, and $RR_i(C_{tr})$ represents the relative risk of acute exposure health outcomes i, which is a function of the concentration of major pollutants C_{tr}.

Here, CREM-HE uses a classic exponential model where β_i is the percentage incidence or mortality change for every 10 μg/m³ increase of $PM_{2.5}$. F_{itr} is the benchmark incidence, and is the number of exposed individuals. CREM-HE mainly describes five types of acute exposure: acute exposure death (AM), respiratory disease hospitalization (RHA), cardiovascular and cerebrovascular disease hospitalization (CHA), emergency (ERV), and bronchitis (CB). The RRs for acute exposure are obtained based on the previous indigenous studies in the PRC of $PM_{2.5}$ health effects. The RRs are shown in Table A2.6.

The following four chronic exposure deaths are described in CREM-HE: chronic obstructive pulmonary disease (COPD), ischemic heart disease (IHD), lung cancer (LC), and stroke. For chronic exposure, CREM-HE applies Equations B-8–B-10 unlike the traditional methods of static analysis for specific years.

Table A2.6: Relative Risk of Acute Exposure to PM$_{2.5}$

Health outcomes	Pollutants	Place	RR (95% Confidence Interval)	References
Acute exposure death	PM$_{2.5}$	Pearl River Delta	1.0042	Xie et al.
			(1.0003, 1.0081)	
Respiratory disease hospitalization	PM$_{2.5}$	Beijing	1.022	Li et al.
			(1.013, 1.032)	
Cardiovascular and cerebrovascular disease hospitalization	PM$_{2.5}$	Hong Kong, China	1.013	Qiu et al.
			(1.007, 1.019)	
Emergency treatment	PM$_{2.5}$	Beijing	1.015	Wang et al.
			(1.011, 1.019)	
Bronchitis	PM$_{2.5}$	Beijing	1.029	Li et al.
			(1.014, 1.044)	

Source: Various publications compiled by authors.

In Equations B-8–B-10, a new subscript "j" is used to represent the different relative risks of death in different age groups. D_{jtr} stands for the population share of age group j in time t in region r. F_{ijrt} and F_{jrt} represent the mortality of health outcome i and all-cause mortality of chronic exposure, respectively. C_{cf} represents the threshold of PM$_{2.5}$ concentration below which no adverse health effects occur. And the RR functions for chronic exposure in Equations B-9 and B-10 are established based on Burnett et al.

$$Cases_{ijtr}^{CE} = [1 - 1/RR_{ij}(C_{tr})] \cdot F_{ijtr} \cdot P_{tr} \cdot D_{jtr} \tag{B-8}$$

$$RR_{ij}(C_{tr}) = RR_i(C_{tr}) \cdot [(F_{ijrt}/F_{jrt})/(F_{irt}/F_{rt})] \tag{B-9}$$

$$RR_i(C_{tr}) = \begin{cases} 1, C_{tr} < C_{cf} \\ 1 + \alpha\{1 - exp[-\gamma(C_{tr} - C_{cf})^\delta]\}, C_{tr} \geq C_{cf} \end{cases} \tag{B-10}$$

Monetizing Health Loss

The morbidity and mortality cases are computed and then evaluated in monetary terms. For the valuation of health outcomes resulting from acute exposure, the valuation table is constructed based on Zhang (Table A2.7). The cost of illness method is used for evaluating the economic burden of illness based on the increase in the consumption of health care. In this way, RHA, CHA, ERV, and CB are monetized based on the per-capita medical expenditure of these health outcomes.

Table A2.7: The Acute Exposure Health Outcomes Valuation Table for CREM-HE
(in 2012 $)

	RHA	CHA	ERV	CB
BJ	1,121.8	3,519.2	59.1	4,893.6
TJ	817.4	2,564.2	38.5	2,747.4
HE	394.6	1,238.1	28.6	1,201.2
SX	442.1	1,387.0	30.1	1,140.1
NM	456.6	1,432.4	30.3	1,664.5
LN	477.6	1,498.4	34.3	1,908.7
JL	466.5	1,463.6	29.3	1,529.6
HL	462.3	1,450.2	32.3	1,431.1
SH	879.5	2,759.1	43.0	6,374.7
JS	567.9	1,781.7	31.7	2,238.4
ZJ	607.4	1,905.6	31.4	3,069.0
AH	384.9	1,207.6	26.7	1,210.7
FJ	408.3	1,281.0	24.4	1,955.3
JX	358.4	1,124.4	25.6	961.6
SD	431.8	1,354.6	29.5	1,913.3
HA	366.0	1,148.0	21.2	1,056.2
HB	431.8	1,354.8	28.2	1,459.5
HN	387.4	1,215.4	33.0	1,375.6
GD	520.0	1,631.3	27.6	2,562.0
GX	387.7	1,216.2	22.5	969.5
HI	510.8	1,602.5	28.0	1,004.4
CQ	430.7	1,351.3	32.8	1,572.0
SC	385.6	1,209.6	26.2	1,213.6
GZ	299.1	938.3	29.1	800.4
YN	322.8	1,012.8	22.2	869.8
SN	354.9	1,113.4	27.1	1,141.8
GS	312.8	981.3	20.4	808.9
QH	429.2	1,346.4	22.6	952.5
NX	361.7	1,134.7	22.7	1,115.1
XJ	340.6	1,068.5	25.6	965.7

CREM-HE = CREM-Health Effect Module.
Source: Authors.

The cost of AM in the study is determined by calculating the value of a statistic life (VSL). The VSL is a measure of the willingness to pay for a small reduction in the risk of mortality, and it is obtained through surveys using the contingent valuation method.

In the study, two sets of VSL were used in view of the large variation of VSL estimations in previous studies and of the possibility that the adoption of the VSL could substantially affect the evaluation results.

According to the results of a local survey conducted in the PRC by Hoffmann et al. (2017), the VSL in 2006 was approximately $0.23 million. This VSL valuation had to be adjusted with reference to the PRC's income elasticity from 2006 to 2035. The adjusted value, estimated to be $0.42 million (2012 price), was adopted as a lower VSL in the study.

Moreover, the study used the "transfer of benefits" method to estimate the high value of VSL. Specifically, based on the Organisation for Economic Co-operation and Development (2012) method, the study used a transfer elasticity of 0.8 to calculate the PRC's corresponding VSL in 2035. According to this method, the PRC's VSL in 2035 would be about $3.72 million (2012 price).

References

Beijing Municipal Ecological Environment Bureau. (n.d.) *Beijing Environment Statement (2013–2017)*. http://sthjj.beijing.gov.cn/bjhrb/index/xxgk69/sthjlyzwg/1718880/1718881/1718882/index.html.

BP. 2018. *Statistical Review of World Energy*. https://www.bp.com/en/global/corporate/energy-economics/statistical-review-of-world-energy.html.

Cai, S., et al. 2017. The Impact of the "Air Pollution Prevention and Control Action Plan" on $PM_{2.5}$ Concentrations in Jing-Jin-Ji Region during 2012–2020. *Science of the Total Environment*. 580, pp. 197–209.

Department of Ecological Environment of Hebei. 2014–2020. *Hebei Environment Statement (2013–2017)*. http://hbepb.hebei.gov.cn/hjzlzkgb/.

Department of Ecological Environment of Shandong. *Shandong Environment Statement (2013–2017)*. http://xxgk.sdein.gov.cn/xxgkml/hjzkgb/.

Department of Energy Statistics, National Bureau of Statistics. 2020. *China Energy Statistical Yearbook*. Beijing: China Statistics Press.

Fu, X., et al. 2017. Increasing Ammonia Concentrations Reduce the Effectiveness of Particle Pollution Control Achieved via SO_2 and NO_x Emissions Reduction in East China. *Environmental Science & Technology Letters*. 4(6), pp. 221–227.

General Office of the People's Government of Shandong Province. 2016. *Implementation Plan of Resolving Excess Capacity of Iron & Steel Industry in Shandong. (2020-10-06)*. [2016-08-21]. http://www.yishui.gov.cn/info/2687/84107.htm.

Hao, J. 2018. *Strategy on Regional Economic Development and Joint Air Pollution Control in Beijing–Tianjin–Hebei Region*. Beijing: Science Press.

Hoffmann, S., Krupnick, A., and Qin, P. 2017. Building a Set of Internationally Comparable Value of Statistical Life Studies: Estimates of Chinese Willingness to Pay to Reduce Mortality Risk. *Journal of Benefit-Cost Analysis*. 8, pp. 251–289.

IEA. 2016. World Energy Outlook Special Report: Energy and Air Pollution. https://www.iea.org/reports/energy-and-air-pollution.

Karplus, V. J., Rausch, S. & Zhang, D. 2016. Energy Caps: Alternative Climate Policy Instruments for China? *Energy Economics*. 56, pp. 422–431.

Li, M., et al. 2018. Air Quality Co-benefits of Carbon Pricing in China. *Nature Climate Change*. 8(5), p. 398.

Li, N., et al. 2019. Does China's Air Pollution Abatement Policy Matter? An Assessment of the Beijing–Tianjin–Hebei Region based on a Multi-Regional CGE Model. *Energy Policy*. 127, pp. 213–227.

Li, P., et al. 2013. The Acute Effects of Fine Particles on Respiratory Mortality and Morbidity in Beijing, 2004–2009. *Environmental Science and Pollution Research.* 20(9), 6433–6444.

Ministry of Ecology and Environment. 2013. *Detailed Rules for the Implementation of the Action Plan for the Prevention and Control of Air Pollution in Beijing, Tianjin, Hebei and Surrounding Areas.* http://www.mee.gov.cn/gkml/hbb/bwj/201309/t20130918_260414.htm.

Ministry of Ecology and Environment. 2021. *Report on the State of the Ecology and Environment in China 2020.* https://www.mee.gov.cn/hjzl/sthjzk/zghjzkgb/.

Ministry of Ecology and Environment. *Notice on the Work Programme on the Full Implementation of Ultra-Low Emissions and Energy-Saving Transformation of Coal-Fired Power Plants* [EB/OL]. (2015-12-11) [2019-03-24]. http://www.mee.gov.cn/gkml/hbb/bwj/201512/t20151215_319170.htm.

Ministry of Ecology and Environment. 2017. Liu Binjiang's Speech at the *Seminar on Clean Development and Environmental Impact of Coal and Electricity in China.* http://www.china-nengyuan.com/news/114502.html.

Ministry of Finance, et al. 2013. *Notice on Continuing to Launch the Promotion and Application of New Energy Vehicles.* http://www.gov.cn/zwgk/2013-09/17/content_2490108.htm.

Ministry of Finance, et al. 2014. *Notice on Further Promoting and Deploying New Energy Vehicles.* http://www.mof.gov.cn/gp/xxgkml/jjjss/201402/t20140208_2512091.htm.

Ministry of Finance, et al. 2015. *Notice on the Financial Support Policy for the Promotion and Deployment of New Energy Vehicles in 2016-2020.* http://www.gov.cn/xinwen/2015-04/29/content_2855040.htm.

National Bureau of Statistics. 2020. *China Annual Statistical Yearbooks.* http://data.stats.gov.cn/index.htm.

National Development and Reform Commission. *Notice on the Action Plan for the Promotion and Transformation of Energy Conservation and Emission Reduction in Coal power (2014–2020).* [EB/OL]. (2014-09-12) [2019-03-24]. http://bgt.ndrc.gov.cn/zcfb/201409/t20140919626242.html.

Organisation for Economic Co-operation and Development (OECD). 2012. *Mortality Risk Valuation in Environment, Health and Transport Policies[M].* Paris: OECD Publishing.

People's Government of Hebei Province. 2014. *Implementation Plan on Resolving Serious Excess Capacity Contradiction.* (2020-10-06) [2014-02-20]. http://info.hebei.gov.cn//eportal/ui?pageId=6809997&articleKey=6822917&columnId=6812848.

Qi, T., et al. 2016. An Analysis of China's Climate Policy Using the China-in-Global Energy Model. *Economic Modelling.* 52, Part B, pp. 650–660.

Springmann, M., Zhang, D. & Karplus, V. J. 2015. Consumption-Based Adjustment of Emissions-Intensity Targets: An Economic Analysis for China's Provinces. *Environmental and Resource Economics.* 61, pp. 615–640.

State Council. (n.d.) *Notice on the Three-year Action Plan to Win the Blue Sky Defense War [EB/OL]. (2018-06-27) [2019-03-29]*. http://www.gov.cn/zhengce/content/2018-07/03/content_5303158.htm.

Tianjin Ecology and Environment Bureau. 2014–2020. *Tianjin Environment Statement (2013–2017)*. http://sthj.tj.gov.cn/env/env_quality/the_state_of_the_environment_bulletin/.

Tianjin Municipal People's Government. 2013. *Tianjin Fresh Air Action Programme. (2020-10-06). [2013-09-28]*. http://www.tj.gov.cn/zwgk/szfwj/tjsrmzf/202005/t20200519_2365538.html.

Tsinghua University. 2020. Study on Coverage, Cap Setting, Allowance Allocation Methodologies, and Supplementary Mechanisms. Unpublished internal report.

United Nations Environment Programme. 2019. *A Review of 20 Years' Air Pollution Control in Beijing*. http://www.unenvironment.org/resources/report/review-20-years-air-pollution-control-beijing.

United Nations Environment Programme. 2019. *Air Pollution in Asia and the Pacific: Science-Based Solutions*. https://www.ccacoalition.org/en/resources/air-pollution-asia-and-pacific-science-based-solutions-summary-full-report.

United States Environmental Protection Agency (US EPA). (n.d.). CMAQ: *The Community Multiscale Air Quality Modeling System*. https://www.epa.gov/cmaq.

———.2006. *Technical Support Document for the Proposed PM NAAQS Rule: Response Surface Modeling: Office of Air Quality Planning and Standards*. Research Triangle Park, NC, 48.

———.2006. *Technical Support Document for the Proposed Mobile Source Air Toxics Rule: Ozone Modeling; Office of Air Quality Planning and Standards*. Research Triangle Park, NC, 49.

Wang, C., et al. 2019. Structural Decomposition Analysis of Carbon Emissions from Residential Consumption in the Beijing–Tianjin–Hebei Region, China. *Journal of Cleaner Production*. 208, pp. 1357–1364.

Wang, G., et al. 2016. Assessment of Health and Economic Effects by $PM_{2.5}$ Pollution in Beijing: A Combined Exposure–Response and Computable General Equilibrium Analysis. *Environmental Technology*. 2016 Dec;37(24):3131-8, pp. 1–8.

Wang, S. X., et al. 2011. Impact Assessment of Ammonia Emissions on Inorganic Aerosols in East China using Response Surface Modeling Technique. *Environ. Sci. Technol*. 45, pp. 9293–300.

Wang. T. 2019. *Emission Co-mitigation Pathways of CO_2 and Air Pollutants of China*. Tsinghua University. Beijing.

Wu, W., et al. 2017. Assessment of $PM_{2.5}$ Pollution Mitigation due to Emission Reduction from Main Emission Sources in the Beijing–Tianjin–Hebei Region. *Environmental Science*. 38(3), pp. 867–875.

Xie P., et al. 2011. Human Health Impact of Exposure to Airborne Particulate Matter in Pearl River Delta, China. *Water, Air & Soil Pollution*. 215(1–4), pp. 349–363.

Xing. J., et al. 2011. Nonlinear Response of Ozone to Precursor Emission Changes in China: A Modeling Study Using Response Surface Methodology. *Atmospheric Chemistry & Physics.* 11(10), pp. 5027–5044.

Xing, J., et al. 2017. BaCAS: An Overview of the Air Pollution Control Cost-Benefit and Attainment Assessment System and Its Application in China. *Magazine for Environmental Managers—Air & Waste Management Association.* April.

Xing, J. 2011. *Study on the Nonlinear Responses of Air Quality to Primary Pollutant Emissions.* Doctor Thesis, School of Environment, Beijing, PRC (in Chinese): Tsinghua University.

Xing, J., et al. 2017. Quantifying Nonlinear Multiregional Contributions to Ozone and Fine Particles Using an Updated Response Surface Modeling Technique. *Environmental Science & Technology.* 51(20), pp. 11788–11798.

Xu, Q. 2018. *Report on the Work of Government of Hebei. (2020-10-06) [2018-02-05].* http://www.hebei.gov.cn/hebei/14462058/14471802/14471805/14867265/index.html.

Yan, Q., et al. 2019. Coordinated Development of Thermal Power Generation in Beijing–Tianjin–Hebei Region: Evidence from Decomposition and Scenario Analysis for Carbon Dioxide Emission. *Journal of Cleaner Production.* 232, pp. 1402–1417.

Zhang, D., Springmann, M. & Karplus, V. J. 2016. Equity and Emissions Trading in China. *Climatic Change.* 134, pp. 131–146.

Zhang, G. 2018. *Report on the Work of the Municipal Government of Tianjin. (2020-10-06) [2018-02-02].* http://www.tj.gov.cn/zwgk/zfgzbg/202005/t20200520_2462487.html.

Zhang, Q., et al. 2019. Drivers of Improved $PM_{2.5}$ Air Quality in China from 2013 to 2017. *Proceedings of the National Academy of Sciences of the United States of America.* 116(49), pp. 24463–24469.

Zhang, X., et al. 2017. Socioeconomic Burden of Air Pollution in China: Province-Level Analysis Based on Energy Economic Model. *Energy Economics.* 68, pp. 478–89.

Zhang, X., et al. 2021. China's Medium-Term and Long-Term Low-Carbon Emission Strategic Scenario. In: *The Research Group on China's Long-Term Low-Carbon Development Strategy and Transformation Path. Carbon Neutrality: China's Low-Carbon Development Action Roadmap for 2020–2050.* Beijing: CITIC Press Group.

Zhang, X. 2016. *Development and Application of Regional Energy Emission Air-Quality Climate Health Model (REACH).* Beijing: Tsinghua University.

Zhang, X. H. 2018. *A Computable General Equilibrium Analysis of Coal Cap and Clean Coal Technologies in China.* Beijing: Tsinghua University.

Zhao, B., et al. 2018. Change in Household Fuels Dominates the Decrease in $PM_{2.5}$ Exposure and Premature Mortality in China in 2005–2015. *Proceedings of the National Academy of Sciences of the United States of America.* 115, pp. 12401–12406.

Zhao, B., et al. 2015. Assessing the Nonlinear Response of Fine Particles to Precursor Emissions: Development and Application of an Extended Response Surface Modeling Technique v1.0. *Geoscientific Model Development.* 8, pp. 115-128.

Zhou, J., et al. 2018. Scenario Analysis of Carbon Emissions of Beijing–Tianjin–Hebei. *Energies.* 11 (14896).

CPSIA information can be obtained
at www.ICGtesting.com
Printed in the USA
JSHW070941160523
41776JS00006B/159